T0281198

Principles and Practice Series

CAPNOGRAPHY

Principles and Practice Series

CAPNOGRAPHY

DAVID O'FLAHERTY

*Consultant Anaesthetist, Portiuncula Hospital, Ballinasloe,
County Galway, Ireland*

Series Editors:

C E W HAHN

Nuffield Department of Anaesthesia, Radcliffe Infirmary, Oxford

and

A P ADAMS

Professor of Anaesthesia, Guy's Hospital, London

BMJ
Publishing
Group

© BMJ Publishing Group 1994

All rights reserved. No part of this publication may be reproduced, stored in a retrieval system, or transmitted, in any form or by any means, electronic, mechanical, photocopying, recording and/or otherwise, without prior written permission of the publishers.

First published 1994
by the BMJ Publishing Group, BMA House, Tavistock Square,
London WC1H 9JR

British Library Cataloguing in Publication Data
A catalogue record for this book is available
from the British Library

ISBN 978-0-7279-0796-7

Contents

Preface

The monitoring of expired carbon dioxide concentration is becoming increasingly commonplace in routine anaesthetic practice. Intraoperative accidents, such as inadvertent oesophageal intubation, valve dysfunction, or disconnection from the ventilator, may be accurately detected with capnography. Moreover, the non-invasive measurement of end tidal partial pressure of carbon dioxide can give valuable information on the circulatory and metabolic state of the patient. The purpose of this book is to give an introduction to the basic principles and clinical applications of capnography. The physiological and technical background is discussed and the focus is on examples that show how the clinical applications are relevant to anaesthetic practice.

The book is presented in three parts. The first section deals with the physiology of carbon dioxide as measured by capnography; the second, the technical details of carbon dioxide measurements; and the concluding sections explore the clinical importance of particular capnograms.

I wish to acknowledge my anaesthetic colleague, Dr Ian Surgeon, at Portiuncula Hospital, who read early drafts of the book and offered many valuable suggestions.

David O'Flaherty
Ballinasloe, Galway

1 Carbon dioxide and monitoring

The pace of change in anaesthetic practice has accelerated in the past decade. Epidemiological and medicolegal studies suggest that anaesthetic practice is safer now than ever before, allowing surgical treatment to be offered to older and less healthy patients. Morbidity resulting from anaesthesia is difficult to quantify, but most investigators agree that death from anaesthesia has become a rare event in developed countries. Increased intraoperative monitoring, such as capnometry and pulse oximetry, can claim some of the credit for the improvement in morbidity and mortality figures.[1]

Capnometry is the study of the shape or design of the changing concentrations of carbon dioxide in expired air. The technique gives rapid and detailed information about each breath and is now widely regarded as one of the most useful monitors for use in anaesthesia and intensive care, providing an excellent early warning system. The rapid rise in popularity of CO_2 monitoring in the past five years reflects its value in ensuring patient safety. Potentially dangerous situations are often first detected by CO_2 monitoring, thus allowing adequate time for assessment of other variables and for corrective action. Capnometry has been shown to be effective in the early diagnosis of such adverse events as oesophageal intubation, hypoventilation, and disconnection of the breathing system. Monitoring and studying trends in end tidal CO_2 also gives valuable diagnostic information on the patient during anaesthesia. As well as information on ventilation, it provides clinicians with a non-invasive monitor of systemic metabolism and circulation. Capnometry has wider applications in a variety of other circumstances including teaching and research, although many doctors are not aware of its full clinical value. The association of certain capnographic patterns with specific circumstances is now recognised and the curves are often diagnostic; indeed the effect of drugs such as narcotics and different malfunctions of breathing systems produce their own "signature" capnograms.

Specific benefits of CO_2 monitoring in anaesthetic and intensive care practice include the ability to verify rapidly positions of the tracheal tube at intubation that could cause considerable morbidity (for example, oesophageal intubation and disconnection). The routine use of capnometry during

1

anaesthesia may give an earlier warning of imminent arterial oxygen desaturation than is possible with pulse oximetry.

Capnometry is a useful, objective, continuous, and non-invasive indicator of adequacy of ventilation. Disturbances in gas exchange, circulation, and metabolism can also be recognised easily. Capnometry has an important role in preventing ventilation mishaps that could result in hypoxaemia, and it is now accepted that the introduction of capnometry and pulse oximetry has prevented many avoidable anaesthetic mishaps. The routine use of capnometry and pulse oximetry greatly reduces the need for frequent arterial blood gas analysis in intensive care.

Terminology

The capnogram (derived from the Greek *kapnos* "smoke" and *graphein* "to write") is a plot of the concentrations of carbon dioxide as a function of time. The device that continuously records and displays the CO_2 concentration in the form of a capnogram wave form is called a capnograph. The capnometer is an instrument for measuring the numerical concentration of carbon dioxide. By definition, capnometers need not all generate capnograms, but all capnographs are, or are a part of, capnometers.

The normal end tidal presence of carbon dioxide is 5.1 kPa (38 mm Hg) at 101.3 kPa (760 mm Hg) barometric pressure (a 5% concentration), and normal arterial values for carbon dioxide tension lie between 4.8 and 5.8 kPa (36 and 44 mm Hg). Hypercapnia or hypercarbia is an abnormally high concentration of carbon dioxide, and hypocapnia (hypocarbia) refers to an abnormally low concentration of carbon dioxide, either in the bloodstream or in exhaled gas as measured by capnometry.

The end tidal CO_2 ($ETCO_2$) concentration is so called because it is the CO_2 concentration measured at the end of the tidal volume expired. $ETCO_2$ very closely approximates the alveolar concentration of CO_2 because end tidal gas is virtually pure alveolar gas.

Historical background

In the early 1620s the Belgian chemist van Helmont made and first described carbon dioxide. Joseph Black in his MD thesis from Glasgow in 1754 proved that carbon dioxide was exhaled during respiration. Until 1824 it was thought that inhaling CO_2 would be fatal.

Yandell Henderson in the United States in 1925 and JS Haldane in England in 1926 demonstrated the physiological significance of carbon dioxide. Using a 70 cubic foot chamber, Haldane observed that when air was rebreathed the resulting increase in respiratory rate was the consequence of carbon dioxide excess. He used a large rubber hose to collect expired gas samples in order to estimate CO_2 tensions in alveolar gas. The

2

principle of Haldane's method of gas analysis was to draw a sample of gas into a graduated burette by raising and lowering a reservoir of mercury held underneath, recording the volume of the gas. Absorption of CO_2 was caused by exposing the gas to sodium or potassium hydroxide; after this the measurement was repeated. The volume of CO_2 in the original sample is simply given by the ratio of the change in volume divided by the total volume, multiplied by 100.

Henry Hill Hickman mixed carbon dioxide with oxygen and showed he could anaesthetise animals safely and reversibly. Carbon dioxide was used in Wisconsin with narcotics as a general anaesthetic in humans in 1928, but the patients convulsed during recovery and consequently CO_2 never found a role in the maintenance of general anaesthesia. Subsequent studies suggested that the convulsions were a result of cerebral vasoconstriction as CO_2 tension fell after brain HCO_3^- had increased during high CO_2 tension.

Luft developed the principle of capnometry during the second world war on the basis of the observation that CO_2 is one of the gases that absorbs infrared radiation. In 1943 he patented two infrared systems that had a practical clinical application. The technique was initially used in rubber manufacture and for environmental monitoring in submarines.

In the early 1950s Elam and Liston introduced the concept of monitoring expired CO_2 during anaesthesia by infrared light absorption. Infrared radiation is absorbed by all gases with more than two atoms in the molecule. If there are only two atoms, absorption occurs only if the two atoms are dissimilar.

The early use of capnometry was limited to research and the technique was not used in clinical medicine for nearly 20 years because of the bulk, weight, and expense of the early monitors. Smallhout and Kalenda illustrated the considerable benefits of capnometry in Europe in the early 1980s, and capnometry is now accepted as a routine part of care in the operating theatre. Technical improvements in solid state electronics during the past decade have made the capnometer one of the most important and widely used monitors in anaesthesia.

A recent development in capnometry is the Danish capnographic method of photoacoustic spectrography. The only detector in this exciting new system is a microphone in the flowing sample stream, detecting the sounds produced by pulsing light of varying infrared wavelengths, each causing the sample to expand with each pulse and creating sound whose frequencies are characteristic of the pulse rates of the infrared light, each tuned to a particular gas. The sample detection system has advantages over conventional systems in that it is simpler, smaller, more linear, and less inclined to drift.[2]

Mass spectrometer systems, based on the separation of gases and vapours by their molecular weights, can sequentially sample all the inspired and end tidal gases and are widely used in operating rooms in the United States.

These systems can monitor a single patient or be used to monitor several patients sequentially in different locations in a theatre suite.

Generating a capnogram

Carbon dioxide concentrations in inhaled and exhaled gases can be presented in either analogue or digital format. The analogue display (capnogram) shows all changes in carbon dioxide concentrations during the inspiratory and expiratory cycles; this requires almost continuous measurements of carbon dioxide in the respired gas and almost continuous plotting of the measured values. The digital format displays the highest and the lowest concentrations of carbon dioxide as detected by the capnometer and reports these values as end tidal and inspired minimal carbon dioxide concentrations, respectively.

The analogue presentation of the CO_2 waveform obtained from breathing systems used in anaesthesia and intensive care is more valuable than a meter or even a fast digital display. Consequently, analogue presentation has superseded digital display in anaesthetic practice, since the breath by breath wave form needs to be displayed to permit the clinician to recognise trends of rising or falling concentrations of carbon dioxide in the inhaled and the exhaled gas. It is much easier to discern such trends while observing a capnographic plot than with rows of discrete figures. Knowledge of the baseline also is important as a rise in the baseline may indicate a fault in some part of the breathing system.

Errors in diagnosis can result when the capnogram is not or cannot be inspected. The availability of the capnographic wave form is essential where fractional rebreathing techniques are employed, as with Mapleson D/E/F and Bain breathing systems. When these systems are in operation a digital display cannot indicate the CO_2 concentration of the end tidal CO_2 plateau as CO_2 also appears in the inspiratory part of the respiratory cycle. Asking the anaesthetist to interpret a digital CO_2 display would be equivalent to requesting a cardiologist to rely solely on a numerical value for heart rate when commenting on the electrocardiogram.

Since CO_2 measurements indicate such a wide range of variables simultaneously, algorithms are now being designed that can distinguish artefacts from clinical abnormalities in capnograms and give rapid automatic interpretations of the different patterns of capnographic wave forms. These advances will help anaesthetists in their inspection of the capnogram and further increase patient safety.

It is important to remember that the capnogram does not reflect gas flow. As outlined in chapter 4, the final part of the capnogram (just before the plateau drops) is often not coterminous with the end of expiration.

4

Similarly, expiration starts with the exhalation of gas from the anatomical dead space, which is identical to fresh gas; however, the capnogram fails to delineate the reversal in flow and registers expiration, with the arrival of alveolar gas, with an almost vertical line. A flow monitor is needed to identify precisely the distinction between inspiration and expiration; consequently capnometry gives information on CO_2 concentration and respiratory frequency but not on the inspiratory-expiratory ratio. Thus the concentration of carbon dioxide measured at the end of the capnographic plateau may not be truly "end tidal." After a healthy person takes a deep breath, the peak exhalation value of carbon dioxide may come very close to the arterial value.

Although $ETCO_2$, which is measured close to the patient's mouth or trachea, closely approximates the alveolar CO_2 concentration, in some circumstances it may be significantly different. For instance, the highest concentration of carbon dioxide in the expired gas may not represent the alveolar or arterial concentration. The highest value is correctly referred to as the peak concentration in the expired gas or, assuming the the peak will occur at the end of expiration, the end tidal concentration.

The closer to the alveolus that gas is sampled, the more faithfully the capnogram should reflect what is in the alveolus and hence in the pulmonary venous blood draining the lung and in the arterial blood. However, this more invasive sampling is not always practical. Intratracheal sampling is hindered by the presence of mucus and moisture in the upper airways.

As already stated, expired carbon dioxide may be continuously measured by mass spectrometry or infrared analysis. Monitoring equipment based on infrared analysis is the most widely used and most cost effective—its measurements are generally taken as synonymous with "capnography." There are two main types of capnographic infrared analysers: mainstream and sidestream. Both types sample gas close to the patient's mouth. Many different models are commercially available and many are suitable, reliable, and reasonably priced. The trend in monitoring is towards a central integrated monitoring unit which incorporates various monitors, each of which can easily be removed for servicing or repairs, and replaced. The standard sizes allow the modules to be readily adapted to the individual needs of the users, and the capnometer is now considered by manufacturers and users alike as a basic component of these newer anaesthetic monitoring facilities.

The magnetic sector and the quadrupole analysers are the two types of mass spectrometers that have become important in anaesthetic practice. They both may act as a dedicated instrument providing continuous monitoring of single patients or as a shared instrument (multiplexed) providing intermittent monitoring of several patients in sequence. A detailed discussion of sampling technology and instruments can be found in chapter 4.

5

1 Tuman K. Evaluating anesthetic outcome: In Stoelting R, Barash P, Gallagher T, eds. *Advances in anesthesia*.St Louis: Mosby, 1991:311–33.
2 Severinghaus J. What's new with CO_2? *Acta Anaesthesiol Scand* 1990;**34**:13–7.

Further reading

Bhavani-Shankar K, Moseley H, Kumar A, Delph Y. Capnometry and anaesthesia. *Can J Anaesth* 1992;**39**:617–32.
Dueck R. Respiratory monitoring. *Current Opinion in Anaesthesiology* 1993;**6**:946–51.
Tobin M. *Respiratory monitoring*. London: Churchill Livingstone, 1991. (*Contemporary Management in Critical Care.*)
Tremper KK, Barker SJ. Fundamental principles of monitoring instrumentation. In: Miller RD, ed. *Anesthesia* Vol 1. 3rd ed. New York: Churchill Livingstone, 1990:957–99.

2 Basic concepts of carbon dioxide homoeostasis

Regulation of breathing

The volume and frequency of ventilation is governed by the respiratory centre in the medulla oblongata. Respiratory homoeostasis is determined in part by neural input from two different types of chemoreceptors in the body; central receptors close by the respiratory centre in the medulla and peripheral receptors in the carotid arteries and aortic arch. These receptors are composed of sensory nerve terminals, which respond to stimuli of various kinds.

Central chemoreceptor activity depends mainly on the carbon dioxide concentration in the blood, expressed as $PaCO_2$ (the partial pressure of carbon dioxide). $PaCO_2$ affects the carbon dioxide concentration and thereby the pH value in the cerebrospinal fluid surrounding the brain and spinal cord. The pH of the cerebrospinal fluid has a direct effect on the respiratory centre: a low pH (high CO_2 concentration) stimulates breathing and a high pH (low CO_2) diminishes breathing. The peripheral receptors are similarly influenced by the pH of the blood: a low pH stimulates breathing and a high pH diminishes it. During spontaneous respiration, respiratory drive automatically ensures that blood $PaCO_2$ is maintained in a narrow range around 5.3 kPa (40 mm Hg). This is achieved by simple adjustments of the respiratory rate and tidal volume. If more CO_2 is produced during exercise, for example, the rate and depth of breathing is increased to compensate. In patients with chronic lung disease the respiratory centre develops tolerance to raised $PaCO_2$ (low pH in cerebrospinal fluid) over a long period of time. Respiratory drive is governed in this circumstance by the oxygen concentration in the blood, expressed as PaO_2 (the partial pressure of oxygen) via the peripheral receptors. When PaO_2 falls to about 8 kPa the respiratory centre is stimulated.

Excitatory and inhibitory neurotransmitters are involved in generating respiratory rhythm and in interpreting chemoreceptor drive. Ventilatory depression can occur by suppression of antagonism of endogenous excitatory neurotransmitters such as N-methyl-D-aspartate in the ventrolateral medulla, as well as by agonist action on inhibitory sites. Acetylcholine, γ-aminobutyric acid, endorphins, and catecholamines are other neurotransmitters that are involved in ventilatory control neural pathways and in maintaining ventilatory drive. Anaesthetic and analgesic drugs can and do interact with many of these transmitter systems in the brain stem in ways that have not been completely elucidated.

Factors that control breathing in the awake state differ from those that influence breathing during sleep, sedation, or anaesthesia. The wakefulness drive seems to predominate over chemoreceptor influences in the awake state. These chemoreceptor drives due to hypoxia and hypercapnia are considered more as emergency reflexes than the factors that control normal breathing. In contrast, during non-rapid eye movement sleep and anaesthesia the chemoreflexes are vitally important in maintaining ventilation, and removal of their influence can lead to apnoea. Ventilatory control during the transitions between awake, sedated, and sleep states can be extremely variable and may contribute to the desaturations seen during recovery from anaesthesia in apparently healthy patients.

Alveolar ventilation

An understanding of normal and diseased patterns of CO_2 elimination is required to interpret correctly and use the information available from capnometry. The efficacy of ventilation can be described by theoretically dividing ventilation into two components: the volume in which perfect gas exchange has taken place (alveolar ventilation) and the volume in which no gas exchange has taken place (physiological dead space). To obtain the alveolar tidal volume, where gas exchange can take place, all the dead space must be subtracted from the tidal volume. Alveolar minute ventilation is obtained by multiplying the respiration rate by the alveolar tidal volume. Causes of insufficient alveolar ventilation include depression of the respiratory centre, paralysis of respiratory muscles as a result of muscular disease or muscle relaxants, increased dead space (for example, in emphysema or pulmonary embolus), or an incorrectly set ventilator.

Dead space

Physiological dead space

The physiological dead space is the sum of the anatomical (airways) dead space and the alveolar dead space. The total dead space in a ventilated patient also incorporates that part of the tidal volume of gas that the patient breathes in that never reaches the alveoli but remains in the mechanical dead space (the tracheal tube, airway adapters, and Y piece) where no gas exchange can take place. For the ventilated patient, this apparatus dead space is treated as if it were a simple extension of the patient's anatomical

8

dead space, and this approach is valid for uncomplicated breathing equipment. The physiological dead space is about 2 ml/kg body weight or 80 ml/ m^2 body surface.

The volume of gas that stays in conducting airways (trachea, bronchial tree, etc) is called the anatomical dead space. Anatomical dead space usually consists of the upper airway and the portions of the bronchial tree that do not have the capability to exchange gas. This volume of air in the anatomical dead space is generally free of CO_2 at the end of inspiration, so the composition of gas here (at this point in the respiratory cycle) is similar or nearly similar to the composition of atmospheric air. The converse is true at the end of expiration, when the anatomical dead space is filled with end-expired alveolar gas. Generally, the normal amount of anatomical dead space may be calculated as about 2 ml/kg of body weight. The anatomical dead space depends on the age, height, and weight of the individual. Other factors influencing the anatomical dead space include lung volume, tidal volume, and whether or not the patient is intubated.

Alveolar dead space

Inspired air can be delivered to areas beyond the anatomical dead space where gas exchange is either incomplete or does not occur. The alveolar dead space is defined as the volume of gas reaching alveoli but not undergoing gas exchange—for example, if the alveoli are poorly perfused or non-perfused. The alveolar dead space is minimal in spontaneously breathing healthy individuals, and under normal conditions it is usually too small to be measured. In certain lung diseases, however, alveolar dead space can become large enough to affect the adequacy of gas exchange. This condition exists when lung units are ventilated but are not perfused. Gas exchange will not take place there and the alveolar gas composition remains similar or nearly similar to that in atmospheric air. Ventilation-perfusion disturbances can become severe enough that the underperfused part of the lung dilutes the CO_2-rich alveolar gas coming from the rest of the lungs, lowering the total end tidal CO_2 in exhaled air; then the level of CO_2 in blood can increase. In extreme cases these affected alveoli will contribute exhaled gas virtually free of CO_2 if no exchange has taken place.

An abnormally large alveolar dead space may be caused by pulmonary thromboembolus and pulmonary hypoperfusion. Some dead space ventilation occurs in the standing position compared to the supine position, particularly in subjects who have low blood pressures. During positive pressure ventilation, alveolar dead space increases appreciably (even in healthy individuals) and can amount to a large proportion of the alveolar ventilation. This effect is caused by the raised intrathoracic pressure inducing a ventilation-perfusion imbalance. Anaesthetic agents and pre-existing lung disease can further increase this effect. Any condition that

9

Influences on CO_2 production

Increased production
- Hyperthyroidism
- Addison's Disease
- Salicylate poisoning
- Malignant hyperpyrexia

Decreased production
- Hypothyroidism
- Hypothermia
- Reye's syndrome

---

inhibits normal blood flow in the lung will result in ventilation without gas exchange.

Production of carbon dioxide

Carbon dioxide (CO_2) is a waste product of aerobic metabolism and is produced by all the cells in all the tissues of the body. The mitochondria produce the major portion of all endogenous CO_2, and the basic metabolic reaction involved is the oxidation of glucose to produce energy, which results in the need to eliminate the byproduct, CO_2. Diseases that influence the consumption of oxygen will also govern CO_2 production (box). Increased production of CO_2 is a feature of hyperthyroidism, Addison's disease, salicylate poisoning, and malignant hyperpyrexia. Decreased production occurs in hypothyroidism, cyanide poisoning, and Reye's syndrome. Changes in metabolic rate will produce proportionate changes in alveolar CO_2 concentration unless a change in alveolar ventilation occurs; this interaction may help in diagnosis if CO_2 is continuously monitored. Hence any condition that alters the basal metabolic rate will also alter carbon dioxide production. Changes in temperature are well recognised as causing changes in oxygen consumption and thus in CO_2 production. Hypothermia substantially decreases metabolism, whereas fever and shivering greatly increase oxygen consumption and CO_2 production.

As the byproduct, CO_2 must be continually removed from the cells. This is achieved by a concentration gradient that allows carbon dioxide to diffuse into the capillary blood. When carbon dioxide is formed by cells involved in the metabolic processes, the CO_2 tension in local tissue will increase above the arterial blood tension. Carbon dioxide molecules will therefore diffuse from the tissues into the plasma of capillary blood.

Once in the capillary blood, the waste CO_2 is transported in the venous circulation from the periphery to the right side of the heart and then in the mixed venous blood through the pulmonary circulation to the lungs for gas exchange. How much CO_2 reaches the alveoli depends on the amount produced during metabolism and on the adequacy of transport to and through the lungs. The very thin membranes (0.2 μm) of the alveoli facilitate gas diffusion between pulmonary blood and the alveolar gas space. Carbon dioxide diffuses into the alveolar gas space down a concentration gradient, and spontaneous respiration or mechanical ventilation continuously keeps the CO_2 concentration lower in the alveoli than in the pulmonary circulation.

The exhalation phase of breathing serves to eliminate CO_2 from the body. Although the lungs are by no means emptied, the final part of the exhaled gas is almost pure alveolar gas. The elimination of CO_2 depends on the condition of the lungs and airways and on the functioning of an integrated respiratory system, both centrally and peripherally. The concentration of CO_2 in the alveoli reflects the balance between the rate of CO_2 production (VCO_2) and alveolar ventilation (VA). Consequently, the measurement of changing CO_2 concentrations is of value for recognising abnormalities of metabolism, ventilation, and circulation, and these three factors are interdependent in changes in $ETCO_2$. Normally, alveolar CO_2 concentration ($PACO_2$) is maintained within a narrow range independent of the metabolic state or the size of the physiological dead space.

Transport of carbon dioxide

Although the main function of the cardiopulmonary system is to oxygenate mixed venous blood so that cells can be adequately supplied with oxygen from arterial blood, blood flow has other important functions. One of these is the removal of waste products of metabolism, such as carbon dioxide, which is carried by venous blood to the lungs, where it is eliminated in the expired gas. Once produced, CO_2 diffuses from tissue cells into venous blood and is carried in three ways: physically dissolved in plasma, bound to proteins, and as bicarbonate ion after chemical conversion in the red blood cell.

CO_2 in plasma

About 5–10% of the CO_2 moving from tissues to lungs travels as dissolved gas in plasma. This is the portion of blood CO_2 referred to as PCO_2 in blood gases and may be considered a general reflection of total CO_2.

Carbon dioxide reacts with water to form carbonic acid (H_2CO_3); this is a slow chemical reaction which requires several minutes to attain equilibrium. Most of the carbon dioxide involved in this reaction remains in the dissociated form. The amount of gas carried in solution is controlled by Henry's law, which states that the mass of a slightly soluble gas that can

11

dissolve in a given mass of a liquid at a specified temperature is very nearly directly proportional to the partial pressure of that gas, provided there is no chemical reaction between the gas and the solvent. The body temperature affects the solubility of gases in liquids—for instance, the solubility of carbon dioxide gas at 44°C is much lower than at 20°C. Therefore in this reaction the carbonic acid concentration is proportional to the PCO_2 in the water since the concentration of dissolved carbon dioxide is proportional to its partial pressure.

Dissolved carbon dioxide amounts to about 1.2 mmol/l in the plasma. A small amount of carbonic acid is formed from the reaction between carbon dioxide, and the concentration of carbonic acid is approximately 0.0017 mmol/l. Also a small quantity of carbon dioxide can combine directly with proteins to form carbamino compounds. Hydrogen ions are produced in this reaction and are buffered in the plasma.

Red blood cells and bicarbonate ions

The body cannot depend solely on the transport of dissolved gas to the lungs because of the limited capacity of this system, especially in situations of increased CO_2 production such as exercise or shivering. The red blood cell is an important transport system for undissolved carbon dioxide. As already outlined, hydration of carbon dioxide to carbonic acid in water alone proceeds slowly, but when carbon dioxide gas enters the red blood cell the effect is dramatic.

Carbonic acid also has a major role in the carriage of CO_2 in red blood cells because the enzyme carbonic anhydrase is present in erythrocytes in high concentrations. This catalyst enormously accelerates the hydration of carbon dioxide to carbonic acid ($CO_2 + H_2O = H_2CO_3$). Almost immediately some of the carbonic acid dissociates into hydrogen and bicarbonate ions, with the hydrogen ions being buffered by the haemoglobin. Ionic equilibrium is maintained within the red blood cell by displacement of the chloride ion from the cell to the plasma by bicarbonate: the chloride (Hamberger) shift. In the presence of a carbonic anhydrase inhibitor such as acetazolamide, carbon dioxide in blood increases and the production of bicarbonate decreases.

The partial pressure of carbon dioxide (PCO_2) in interstitial fluid surrounding cells is similar to that inside the cell because gas can move so easily from one compartment to the other, and under normal circumstances this PCO_2 is 6.1 kPa (46 mm Hg). Venous blood serving the capillaries draining the interstitial areas thus also has a PCO_2 of about 6.1 kPa (46 mm Hg). Arterial blood with a PCO_2 normally of 5.3 kPa (40 mm Hg) supplies the interstitial space and the cells. The small intracellular-arterial

12

gradient facilitates carbon dioxide transfer from cell to blood via the interstitial fluid.

Alveolar-arterial CO_2 tension difference

As outlined previously, end tidal carbon dioxide concentration is the partial pressure of CO_2 at the end of an exhaled breath and closely reflects $PaCO_2$, providing that alveoli from all parts of the lung are emptying synchronously, thus generating a capnogram pattern with an almost horizontal plateau. The normal tension difference between alveolar and arterial CO_2, $P(a-A)CO_2$, which is defined as the arterial to alveolar difference for CO_2, is usually expressed in terms of partial pressure. The normal arterial pressure for CO_2 ($PaCO_2$) is about 5.3 kPa (40 mm Hg) and is a reflection of the CO_2 gas molecules in the plasma of arterial blood. The normal alveolar pressure for CO_2 ($PaCO_2$) is also about 5.3 kPa (40 mm Hg) and is generally a composite value from all alveoli participating in ventilation. Consequently, the ideal difference between $PaCO_2$ and $PaCO_2$ under normal conditions is zero.

Although $PaCO_2$ (or $PETCO_2$, which approximates $PaCO_2$) closely mirrors the concentration of CO_2 in blood as determined by arterial blood gas analysis, the values are not exactly the same. Normally, blood leaving ventilated alveoli mixes with blood from parenchymal lung tissue and also with blood passing through non-ventilated alveoli, creating a venous admixture. This venous admixture accounts for the normal alveolar-arterial CO_2 tension difference, $P(a-ET)CO_2$, which varies from 0.3 to 0.6 kPa (2 to 5 mm Hg), with end tidal carbon dioxide concentration concentration lower than the arterial value. The difference is useful because it gives additional information on the patient—it can be considered as an index of alveolar dead space, and significant changes should be investigated clinically.

The normal alveolar-arterial CO_2 tension difference is much greater in disease states and least when patients' lungs are mechanically ventilated with large tidal volumes. When ventilation and perfusion are mismatched, significant differences in alveolar-arterial CO_2 tension ($P(a-ET)CO_2$) occur, generally as a result of blood that does not exchange gas. End tidal values will also be affected. The inability of blood to exchange gas at the lungs occurs primarily for two reasons: blood is not adequately presented to the lungs to allow gas exchange to occur, or gas is not exchanged at the lungs because alveoli are not adequately ventilated. The $P(a-ET)CO_2$ value depends on the alveolar dead space, which in turn is mainly influenced by the relative distribution of ventilation and perfusion (V/Q) within the lungs. Changes in either of these variables—for example, in pulmonary disorders (emphysema), pulmonary embolism, decreasing cardiac output,

13

or hypovolaemia—can induce variations in $P(a-ET)CO_2$. In the anaesthetised patient, ventilation and perfusion become slightly mismatched, but this is not clinically important. The $P(a-ET)CO_2$ difference varies from patient to patient; it is lower in infants and smaller children and increases with age. It decreases with large tidal volumes and low frequency ventilation.

Negative $P(a-ET)CO_2$ differences

The end tidal value for CO_2 may occasionally be greater than the arterial value. This situation has been reported to occur in critically ill children and also in pregnant women up to two weeks post partum.

In each pregnancy the $P(a-ET)CO_2$ difference is small, and $PETCO_2$ approaches $PaCO_2$ at the end of first trimester. This probably reflects the many physiological changes associated with pregnancy, which are apparent by 12 weeks of gestation. There is a 25–30% increase in the cardiac output at the end of first trimester over the levels before pregnancy. Cardiac output rises to a plateau of about 1.5 l/min above the average postpartum level well before the end of the first trimester.

An increase in blood volume occurs as early as four to six weeks after conception and rises to 15–20% over the pre-pregnancy level at 12 weeks of pregnancy. The plasma volume expands steadily from the first month of pregnancy. After an initial drop, the red cell volume also increases during pregnancy. However, the plasma volume expands proportionately more than the red cell volume, resulting in haemodilution. These factors (increased cardiac output, haemodilution, and increased blood volume) probably result in superior perfusion of the alveoli and improved gas exchange.

The increased cardiac output associated with pregnancy increases the number of alveoli with low ventilation-perfusion ratios (alveoli with high CO_2 concentrations). As pregnancy advances, reduced functional residual capacity and increased CO_2 production result in further increase in alveolar PCO_2, increasing the likelihood of negative $P(a-ET)CO_2$ values occurring in women during pregnancy.

A negative difference can also occur during anaesthesia, possibly owing to better ventilation of dependent, well perfused alveoli during intermittent positive pressure ventilation with large tidal volumes and low respiratory rates. Gas emptying from small alveoli reaching the mouth, which occurs during the respiratory cycle in healthy lungs at large tidal volumes and low frequencies, may also contribute. These circumstances allow the compartments with a low ventilation-perfusion ratio (alveoli with higher CO_2 concentrations) to make a more substantial contribution to gas exchange. Gas would normally have remained in the airways, resulting in small tidal volumes. The overall effect of these factors is that the terminal portion of

14

Phase III of the capnogram exceeds the mean $PaCO_2$, and hence end tidal CO_2 exceeds the mean arterial PCO_2.

$PETCO_2$ as an estimate of $PaCO_2$

In healthy patients, where alterations in $ETCO_2$ can be regarded as indicating changes in $PaCO_2$, capnometry is a useful non-invasive tool to monitor $PaCO_2$ and hence the ventilatory status of patients during anaesthesia or in the intensive care unit. Monitoring end tidal oxygen values did not become as popular as use of end tidal carbon dioxide because of the variable alveolar-arterial gradient. An initial blood gas analysis can establish the $P(a-ET)CO_2$ difference, and further alterations in $PaCO_2$ may be assumed to be parallel with those in $PETCO_2$, thus making repeated arterial puncture unnecessary.

The relation of end tidal and arterial CO_2 concentrations varies during anaesthesia and surgery. The prime cause of a widened difference is greatly increased alveolar dead space ventilation, which occurs if gas exchange takes place in a part of lung that is not well perfused but still well ventilated. The unperfused part of lung tissue will dilute the CO_2-rich alveolar gas coming from the rest of the lungs, lowering end tidal CO_2 and hence resulting in a bigger $P(a-ET)CO_2$. Possible causes of an increased difference are the lateral decubitus position, pulmonary hypoperfusion, and thromboembolism. Several other factors, such as changes in body temperature and cardiopulmonary bypass with continued ventilation, can result in changes in the ventilation-perfusion status of the lungs, making precise predictions of $PaCO_2$ from $PETCO_2$ measurements quite difficult.

The basic disease process in pulmonary embolus, whether the source is blood clots, fat, or air, is an occlusion of blood flow through affected capillaries, resulting in a portion of blood flow that does not exchange gas. Blood may be redirected locally to areas where ventilation is better, and gas exchange will then occur through these areas. The end result of this clinical problem is that a certain percentage of ventilated alveoli contribute CO_2 values far lower than other alveoli. End tidal values are thus lower than arterial CO_2 values and the $P(a-ET)CO_2$ is increased.

Emphysema is characterised by alveolar air trapping. In severe emphysema, functional residual capacity increases because of air trapping, causing an increase in dead space ventilation because gas is not exchanged. Carbon dioxide is not adequately removed from blood and so CO_2 retention occurs and the $P(a-ET)CO_2$ is widened.

A minor degree of dead space ventilation may occur in the standing position in comparison to the supine position, especially in people with low blood pressures. A greater increase in dead space ventilation occurs in severe hypotension, where the uppermost lung zones are poorly perfused because of inadequate pulmonary artery pressure to move blood into these

15

Influences on $P(a-ET)CO_2$ values

- Position of patient
- Hypothermia
- Prolonged anaesthesia
- Known cardiovascular or respiratory diseases

areas of the lungs. The resulting ventilated but non-perfused portions of the lung increase the $P(a-ET)CO_2$.

In these clinical disturbances, which result in increased alveolar dead space and dead space ventilation, ventilation continues despite poor or non-existent perfusion through affected areas. Ventilation-perfusion ratios are high because the amount of ventilation greatly exceeds the amount of perfusion through the affected areas. The arterial-alveolar differences are often much wider than normal since reduced blood flow through these areas does not allow gas exchange and exhaled CO_2 concentrations become lower than normal.

Capillary shunts occur when alveoli that would be normally ventilated are not adequately ventilated. As a consequence, blood flow may reach an affected portion of lung but no gas exchange takes place because of defective or unventilated alveoli. In contrast with dead space ventilation, ventilation-perfusion values approach zero because perfusion far exceeds ventilation. Alveolar ventilation continues through all normal alveoli, which combine to form an average alveolar CO_2 value that is around 5.3 kPa (40 mm Hg). Since mixed venous PCO_2 is normally 6.1 kPa (46 mm Hg), even large amounts of shunted blood do not greatly alter $PaCO_2$ values. Ventilation of normally perfused lung units is often increased to maintain a normal $PaCO_2$. $P(a-ET)CO_2$ does not widen significantly in shunt perfusion even though clinical compromise and severe arterial desaturation may be present. When the shunt fraction increases to over 20% of cardiac output, the associated increase in dead space causes the CO_2 difference to widen. Clinical examples of shunt perfusion include a foreign body or secretions blocking a bronchus, a tracheal tube failing to ventilate the left lung, or severe atelectasis. Capnometry therefore does not highlight shunt perfusion to the same extent as it does dead space; pulse oximetry, which shows desaturation, is more likely to reveal shunt perfusion.

When a steady state of CO_2 has been reached under constant minute ventilation, changes in $ETCO_2$ normally arise from changes in circulation—in which case metabolic or acid base balance changes can be discounted. Except in malignant hyperthermia, metabolic changes during anaesthesia are normally slight and gradual. The enlargement in

16

$P(a-ET)CO_2$ observed after changing the patient's position could be related to the alterations in ventilation-perfusion relationship that are known to occur in the lateral decubitus position. The anaesthetised patient in the lateral decubitus position, with or without paralysis, has a non-dependent lung that is well ventilated but poorly perfused, whereas the dependent lung is well perfused but poorly ventilated. This results in an increased mismatch of ventilation and perfusion. As a result the dead space increases, leading to an enlarged $P(a-ET)CO_2$. Lateral decubitus positioning has been shown to produce an increase of up to 13% in physiological dead space.[1]

Hypothermia can also influence $P(a-ET)CO_2$ values. Because gas solubility, pH, and pK of carbonic acid all change with temperature, there is a substantial physiological decrease in $PaCO_2$ with hypothermia for a given CO_2 content of blood.

Changes in $P(a-ET)CO_2$ could also be related to an impairment in gas exchange occurring with prolongation of anaesthesia. Patients given enflurane and nitrous oxide have appreciable increases in shunt and large increases in atelectasis 30–90 minutes after induction of anaesthesia.

Considerable increases in $P(a-ET)CO_2$ have been reported during laparoscopic procedures with CO_2 insufflation in patients with preoperatively known cardiovascular or respiratory diseases. One study comparing open and closed cholecystectomies found a difference of the mean values before the insufflation of 0.8 kPa (6 mm Hg); after the insufflation the difference was 1.4 kPa (11 mm Hg), with the changes of end tidal carbon dioxide concentration smaller than the changes in $PaCO_2$.[2] Hypercapnia and acidosis were so severe in 20% of patients that it was decided to convert to a standard open cholecystectomy. The end tidal carbon dioxide concentration may therefore be an unreliable indicator of the patient's condition during laparoscopic surgery.

Mechanical ventilation with high airway pressures may cause overextension of alveoli, resulting in an increase in dead space ventilation. This overextension can result in occlusion or compression of adjacent pulmonary capillaries, removing their capability to partake in gas exchange.

In summary, the principal causes of pulmonary dysfunction during anaesthesia are shunt development caused by compression atelectasis and ventilation-perfusion mismatch. The dependency of these factors on the age of the patient with the presence of obstructive lung disease is well recognised.

Elimination of carbon dioxide in the lungs

Mixed venous blood flows into the pulmonary capillary bed with a PCO_2 of 6.1 kPa (46 mm Hg) in humans at rest. As this is higher than the alveolar PCO_2 of 5.3 kPa (40 mm Hg), carbon dioxide leaves the mixed venous blood and diffuses across the alveolar capillary membranes into alveolar gas and is expelled from the lungs as mixed expired gas. Carbon dioxide and hydrogen ions are released from the haemoglobin molecule during the passage through the lungs as the haemoglobin becomes oxygenated. The hydrogen ions combine with bicarbonate ions (bicarbonate shifts back into the red cells) to form carbonic acid. The carbonic acid is rapidly broken down by carbonic anhydrase to form water and carbon dioxide. The carbon dioxide released from the haemoglobin diffuses across the red cell membrane to the plasma and thence to the alveolar gas. Carbon dioxide leaves the red blood cell and the plasma by this mechanism until the PCO_2 in the plasma is equal to that in the alveoli.

Carbon dioxide in anaesthesia

Carbon dioxide has been used less in anaesthesia in the past decade because of increased awareness of the hazards associated with its use, together with the introduction of intravenous induction and relaxant agents. Its main indication was as a stimulant to respiration, especially during a gaseous induction or after a period of artificial hyperventilation. Carbon dioxide was also used controversially to increase cerebral blood flow during carotid artery surgery, even though hypercapnia was known to induce "stealing" of blood away from an ischaemic area of brain. In the 1970s it was common practice to hyperventilate patients to ensure adequate oxygenation and even to deepen anaesthesia. Disadvantages of prolonged hyperventilation included coronary and cerebral vasoconstriction, respiratory alkalosis, and respiratory centre depression, and these factors contributed to a more difficult and prolonged recovery.

Alveolar minute ventilation is usually adjusted nowadays by anaesthetists to achieve normocapnia, where end tidal carbon dioxide concentration is in the range of 4.8–5.7% (4.8–5.7 kPa; 36–43 mm Hg), and the facility to administer carbon dioxide during routine anaesthesia has mostly been removed from newer anaesthetic machines. With normocapnia, any disturbances in ventilation, circulation, and metabolism can be more easily recognised. Unnecessary deviation from this range leads to an imbalance in acid-base homoeostasis. In normocapnic patients spontaneous breathing can be instituted more easily at the end of the anaesthetic, and recovery is usually more rapid.

During anaesthesia the respiratory muscles are paralysed, and it is the task of the anaesthetist to maintain ventilation to a suitable carbon dioxide level concentration. Previously this was estimated by calculating ventilation needs from nomograms, but nowadays using a CO_2 monitor is the easiest objective way of determining ventilation since it is a non-invasive and continuous method.

18

Effects of carbon dioxide

Carbon dioxide

- About 0.3% CO_2 in the atmosphere
- Non-flammable
- Does not support combustion
- Combines readily with water to form carbonic acid
- Molecular weight 44 (same as nitrous oxide)
- Supplied as liquid in grey cylinders at 50 kPa at 15° C
- Laboratory production from fermentation of grain in preparation of alcohol or from combining a strong acid (HCl) with a carbonate
- Commercial production is usually by action of heat on calcium carbonate or magnesium carbonate in preparation of their oxides

---

Hypercapnia

Changes in $PaCO_2$ mainly affect the cardiovascular system. There is a progressive increase in blood pressure, cardiac output, and heart rate owing to indirect sympathetic stimulation with increasing hypercapnia. Prolonged hypercapnia during anaesthesia has undesirable side effects (box). At high $PaCO_2$ (>10 kPa) cardiovascular depression begins to occur as a result of direct myocardial depression. Acute changes in $PaCO_2$ may lead to changes in pH, which may induce potassium shifts and predispose the patient to cardiac arrhythmias. Clinical manifestations of hypercapnia include tachycardia, sweating, hypertension, headache, and restlessness; these are often experienced by patients recovering from anaesthesia or by elderly patients with severe pulmonary disease.

Specific circulations, such as coronary or cerebral, dilate with hypercapnia and constrict in the presence of hypocapnia, and these interactions may be used to advantage in specific circumstances. In neuroanaesthesia, for example, hypocarbia is used deliberately to reduce brain volume and

---

Side effects of prolonged hypercapnia during anaesthesia

- Susceptibility to cardiac arrhythmia with volatile anaesthetic agents
- Increased cardiac output
- Increased intracranial pressure
- Pulmonary vasoconstriction
- Peripheral vasodilatation

19

intracranial pressure. Hyperventilation is purposely induced to improve the operative conditions during the critical phase of the operation for cerebral aneurysm repair. Excessive hypocarbia may, however, be dangerous as the induced vasoconstriction may reduce the supply of nutrients to below the basal requirements of tissues.

Acid-base balance

Carbon dioxide plays a pivotal role in maintaining acid-base balance. Changes in pH or bicarbonate concentrations affect carbon dioxide in blood. Addition of hydrogen ions (pH decrease), either from an external source such as the infusion of hydrochloric acid or from an internal source such as the liberation of lactic acid during severe exercise, cause an imbalance in the carbonic acid reaction. Extra hydrogen ions will cause bicarbonate to combine with hydrogen ions to form carbonic acid, liberating carbon dioxide. The Henderson (and the logarithmic Henderson-Hasselbalch) acid-base formulas show the amount of bicarbonate combining with hydrogen and how much carbon dioxide will be liberated. Acidic conditions increase PCO_2 in the arterial blood, provided compensatory mechanisms are not invoked.

1 Pansard J, Cholley B, Devillers C, Clergue F, Viars P. Variation in arterial to end tidal CO_2 tension differences during anesthesia in the kidney rest lateral decubitus position. *Anesthesia and Analgesia* 1992;75:506–10.
2 Rothen H, Hedenstierna G. Pulmonary gas exchange. *Current Opinion in Anaesthesiology* 1992;5:831–8.

Further reading

Neufeld G. Pulmonary gas transport/capnography. *Current Opinion in Anaesthesiology* 1993;6:956–62.
Nunn JF. *Applied respiratory physiology*. 4th ed. London: Butterworth, 1993.
Ward D. Control of breathing ventilatory failure and sleep apnoea syndromes. *Current Opinion in Anaesthesiology* 1992;5:843–7.

3 Measuring carbon dioxide

Mass spectrometry, infrared analysis, and the new technique of Raman scattering are the main methods used for continuously measuring carbon dioxide. The mass spectrograph separates gases and vapours of differing molecular weights. It is an excellent capnometer and can measure not only carbon dioxide and other physiological gases but also nitrous oxide and other anaesthetic agents. Unfortunately, it is very expensive. Because it is bulky, it is usually situated centrally and is used automatically to monitor several patients. The extensive use of mass spectrometry in Britain and Ireland is precluded by high initial capital expenditure. Raman spectrometry, which uses the principle of Raman scattering for carbon dioxide measurement, has the potential to combine the lower expense of infrared with the multigas capability of mass spectrometry.

Infrared analysis is more commonly used and is less expensive than mass spectrometry. Infrared capnometry uses infrared spectrometry, and this approach is generally considered as synonymous with capnometry. It has some distinct advantages for medical monitoring (box).

Infrared analysis

The infrared spectrum begins just beyond the red part of the visible spectrum; wavelengths range from 1 µm to 40 µm. Infrared rays are given off by all warm objects and are absorbed by gases whose molecules are composed of more than one element. The absorption of infrared energy increases molecular rotation and vibration. Gases that absorb infrared radiation must be composed of molecules that are both asymmetric and polyatomic, such as nitrous oxide. Molecular asymmetry suggests that

---

Advantages of infrared capnometry for medical monitoring

- Concentration of absorbing gas in mixture can be determined reliably from the fall in the intensity of a particular wave length of infrared energy after it has passed through the mixture
- Causes no permanent changes in the molecules exposed
- Sources of infrared energy are readily available
- Transmission materials (windows and filters) readily available

---

different atoms exist in the same molecule and that vibrations induced by infrared energy alter the molecule's dipole moment. The frequency of these vibrations depends on the masses of the atoms as well as the strength of the bond holding the atoms together. The wave length that is absorbed by a particular gas remains constant and specific. These frequencies correspond to frequencies found within the infrared part of the spectrum. Absorption of infrared energy increases the amplitude of the vibration but does not alter its characteristic frequency. Carbon dioxide, nitrous oxide, anaesthetic agents, and water fulfil the criteria of asymmetry and polyatomy and consequently absorb infrared energy, whereas symmetric molecules like helium, argon, hydrogen, oxygen, and nitrogen do not. Specific gases absorb particular wavelengths, producing absorption bands on the infrared electromagnetic spectrum. When infrared radiation is projected through a gas mixture containing an absorbent gas its intensity is diminished, which allows the specific absorption band to be identified for the particular gas.

The infrared region of the spectrum is particularly appropriate for measuring carbon dioxide because carbon dioxide has a strong absorption band at 4.26 μm. Carbon dioxide can be measured with a narrow band pass optical infrared filter, which prevents the passage of radiation that would be absorbed by gases other than CO_2. There is some interference from water vapor and from nitrous oxide, which has the same mass number and an adjacent absorption band (fig 3.1). There is a remarkable similarity between carbon dioxide and nitrous oxide—apart from having the same mass number, both deviate from ideal gas law to the same degree and both are very soluble in water (table).

As already outlined, infrared absorption by carbon dioxide is a measure of vibrational motion of atoms within a molecule of carbon dioxide. These induced vibrations depend on factors both internal and external to the molecule and can alter the shape of the absorption bands. Internally, the vibrational energy attained depends mostly on the bond energies between atoms and their relative atomic weights. When CO_2 molecules vibrate in the presence of nitrous oxide molecules, the collisions affect their vibrational energy states and thus the absorption of infrared light. External forces influence the frequency of vibrations by nature of the spatial arrangement of atoms, intermolecular forces, and collisions between molecules, as well as the ambient pressure. It is by this mechanism of pressure broadening that high concentrations of elemental symmetrical gases such as nitrogen or oxygen that do not have infrared absorption bands and that would not normally be expected to absorb infrared radiation can influence the absorption bands of carbon dioxide and affect the sensitivity of the infrared CO_2 analyser. The result is that the absorption spectrum becomes broader and the degree of overlap in the absorption bands of different gases varies according to the gas concentrations. The resulting overestimation of CO_2 is compensated for electronically—the more sophisticated analysers vary the

22

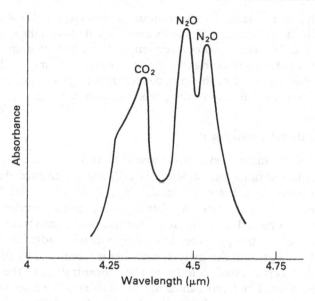

Fig 3.1 Absorbance of infrared radiation by carbon dioxide and nitrous oxide depends on the wavelength of the radiation. (Reproduced with permission from Parbrook GD, Davis PD, Parbrook EO. Basic physics and measurement in anaesthesia. 3rd ed. London: Butterworth Heinemann, 1990)

Physical properties of carbon dioxide and physiologically related gases. (Reproduced with permission from Nunn JF. Respiratory measurements in the presence of nitrous oxide. Br J Anaesth 1958; 30: 254)

	Nitrous oxide	Carbon dioxide	Oxygen	Nitrogen
Molecular weight	44.013	44.007	32.000	28.016
Normal molar volume	22.25	22.25	22.4	22.4
Boiling point (°C)	−89.5	−78.5	−183	−196
Solubility in water at 20°C	0.63	0.878	0.031	0.0164
Viscosity at 20°C (micropoises)	146	146	200	174
Thermal conductivity at 0°C $(J/s/cm^3 \times 10^4/°C)$	1.51	1.45	2.44	2.43
Specific heat at constant pressure (J/g/°C)	0.892	0.898	−	1.030
Velocity of sound (m/s at 0°C)	258	260.3	314.84	333.64

compensation according to the changing concentration of the interfering gases, which they can simultaneously measure.

Similarly, the presence of water vapour can interfere with the estimation of CO_2. Selecting a filter which allows penetration of only the desired wave length is vital for accurate measurement of CO_2, but even an ideal filter does not succeed in eliminating the effects of water vapour entirely. Carbon monoxide poses less of a problem as concentrations generally are low (high concentrations are still only in the parts per million range).

The infrared analyser

Infrared CO_2 analysis systems consist of three basic components: a source of infrared radiation, an analysis cell, and an infrared detector (fig 3.2). An infrared light beam (classically from a hot wire source) is passed through an interference filter to obtain the required wavelengths. The infrared beam is passed through the gas sample in the analysis chamber and is then focused on the photodetector. Any infrared radiation that is not absorbed by the gases in the analysis chamber is monitored by the detector, whose output is processed to indicate the concentration of the gas in the analysis chamber. The filters can be mounted on rotating discs or choppers to produce an alternating signal; this helps reduce drift, which is more likely to occur with a steady signal. The blank interval provided by the chopper also serves to redefine the zero line.

Modern capnometers incorporate a light emitting diode to produce light of the required wavelength. A solid state photodetector (instead of a micromanometer, as in the Luft system) measures the amount of light reaching it alternately through the analysis and reference cells, with the beam chopped 4000 times a minute. The end result is an electrical output consisting of a series of pulses whose height varies with the CO_2 concentration in the analysis cell. This compares with the pulses produced in the detector cell in systems that incorporate a micromanometer. Some designs omit the chopper and the infrared diode is switched on and off by a microprocessor. The infrared cell is the most critical part of the system and must be protected from contamination by liquids or particulate matter as this may lead to an erroneous signal.

The narrower the wavelength filtered, the more specific the output for CO_2. Introducing variable filters allows increased variability of gaseous measurement. When different filters are mounted on the same rotating disc, several sample compounds can be analysed simultaneously.

Stabilising the readings

Practical capnometers have several other components in the infrared system to stabilise the readings and make them less susceptible to disturbances. For example, in the classic double beam Luft infrared (UNOR system) the light source is a broad spectrum emitter that includes the

24

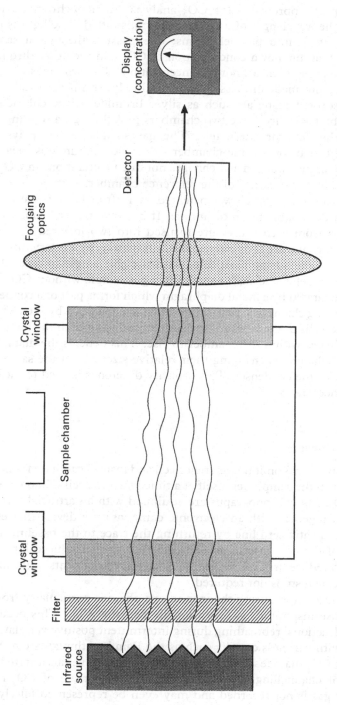

Fig 3.2 The infrared analyser (Reproduced with permission from Parbrook GD, Davis PD, Parbrook EO. Basic physics and measurement in anaesthesia. 3rd ed. London: Butterworth Heinemann, 1990)

absorption peak appropriate for CO_2 analysis. A monochromatic filter allows only the wavelength of interest to be transmitted. The light is then split by a crystal into parallel beams, one passing through a sample chamber with an unknown concentration of CO_2 and the other through a reference chamber with a known concentration. The windows of these chambers are not made of glass (which would absorb infrared radiation) but non-absorbent materials such as silver bromide, silver chloride, or sapphire. The beams from the two chambers pass through a rotating disk and reach the detector alternately. The gas sample to be analysed is circulated through the sample chamber to provide continuous measurement of carbon dioxide, and any changes not due to alterations in CO_2 can be corrected by the presence of the reference chamber.

The URAS system, which was patented by Luft in 1943, is based on a variation on the double beam technique. It has two infrared sources: the infrared rays from each source are focused into two infrared beams by concave mirrors and are then interrupted, alternately, by a rotary disc. One beam crosses a measuring shaft filled with the gas sample to be analysed and the other a reference shaft containing a gas sample without CO_2. The beams are separated by a metal diaphragm which forms part of a condenser in the measuring chambers. The infrared rays are absorbed by the CO_2 and by the black walls of the measuring chamber, inducing both the temperature and pressure within the chamber to rise. This deflects the diaphragm between the chambers, changing their relative sizes and at the same time the capacity of the condenser. This capacity difference is used to measure the CO_2 concentration.

Automatic zeroing

Some instruments omit the reference cell and instead calibrate for zero as reference from the sample cell itself at a time when the cell is known not to contain CO_2. Many capnographs are equipped with an artificial zero line mechanism, together with an electronic compensatory device that establishes an automatic zero line by returning the trace to the baseline at the beginning of the next inspiration. This facility was introduced into older machines to reduce drift caused by the electrical components; it means the chopper mechanism is not required.

Automatic zeroing during the inspiratory phase is particularly troublesome because inspiratory gas occasionally contains CO_2. This occurs in controlled fractional rebreathing during intermittent positive ventilation of the lungs with valveless circuits such as the Bain breathing system. Some inspiratory CO_2 may be due to a failed expiratory valve, a saturated CO_2 absorber, or channelling. In these situations, an increase of CO_2 in the inspiratory gas is not recorded and may even be represented falsely as a

26

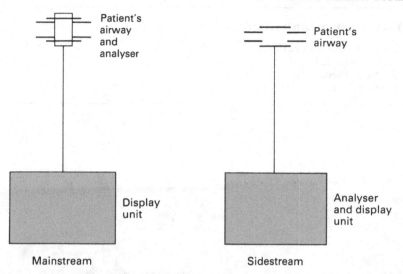

Fig 3.3 *Mainstream and sidestream instruments differ in the position of the analyser relative to the display unit*

drop in $ETCO_2$. Capnographic curves often fail to be recorded on mainstream capnographs with an electronic zero line mechanism when the patient's ventilation rate falls below seven breaths a minute.

Types of capnometer

Mainstream analysers

Mainstream analysers generate a capnogram almost instantaneously since the gas is analysed as it passes through the cuvette containing a specially designed airway adapter that allows the measuring head to be directly interposed into the breathing system. The airway adapter incorporates an analysis cell with an infrared source, detector, and associated electronics.

Sidestream analysers draw a continuous sample of gas from the respiratory circuit and analyse it in an infrared measurement head within the monitor enclosure, away from the patient (fig 3.3). Consequently, the response of the sidestream analyser is preceded by the time necessary for transport of gas from the sample port to the analyser.

The mainstream system offers the advantage of very fast response; any delay in response in mainstream analysers is due to the analyser itself. The sensor in the mainstream analyser is usually inserted between the tracheal tube and the breathing circuit. Conventional mainstream analysers tend to

27

Mainstream capnometers

- Hewlett Packard 47210A
- Novametrix 1260
- Siemens-Elema 930
- Nellcor Ultracap

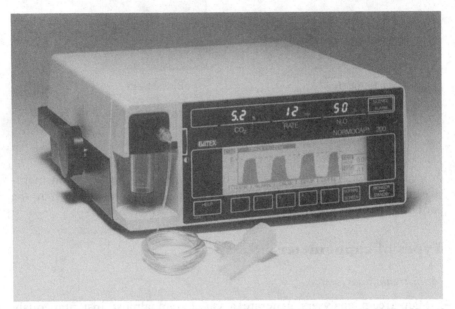

Fig 3.4 The Normocap 200 from Datex. (Datex machines are sold in the United Kingdom by S&W Vickers Ltd, Ruxley Corner, Sidcup, Kent DA14 5BL)

be more bulky than sidestream analysers and have heavy sensors. When connected by means of an electrical cord to the main part of the analyser these sensors can produce traction on the tracheal tube; if unsupported they can cause the tube to kink or dislodge. The sensor heads in the second generation of mainstream analysers are more compact and lighter.

Overall, mainstream analysers tend to be more accurate than sidestream analysers, with fewer associated problems due to the avoidance of the sampling line. Great care needs to be taken of the sensors as they are vulnerable and replacement of the delicate components in the mainstream adaptor is expensive.

28

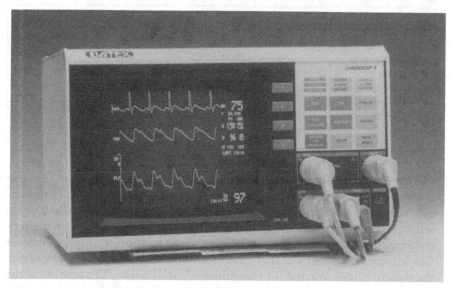

Fig 3.5 The Cardiocap II from Datex

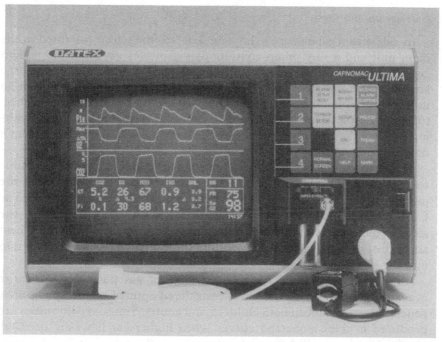

Fig 3.6 The Ultima from Datex

Sidestream capnometers

- Datex Normocap 200
- Cardiocap II
- Ultima
- Puritan Bennett CD-102
- Ohmeda 5200
- Spacelabs 540

The volume of the cuvette in mainstream analysers adds dead space to the system and is a disadvantage when they are used in infants. However, the need for gas sampling and scavenging is eliminated. Mainstream sensors function above body temperature (at about 40°C) to prevent condensation of water vapour, which can lead to spurious high values for CO_2 concentration. The heated cuvette should be carefully kept away from the patient's head to avoid the possibility of facial burns. The heating of the sensor cannot, however, prevent occlusion of the optical windows of the cuvette by secretions or nebulisation that absorb infrared radiation and lead to a falsely high CO_2 concentration.

Sidestream analysers

In sidestream capnometers (box; fig 3.4–3.6) the sensor is located in the main unit and a tiny pump continuously aspirates gas from the patient breathing circuit and pumps it along the fine catheter leading to the analyser. The sample is drawn, typically at a rate of 150 ml/min from an inexpensive, easy to connect, lightweight connector inserted at the tracheal tube or anaesthesia mask connector. In unintubated, spontaneously breathing patients the tip of the sampling tube can be inserted into the patient's nostril.

The airway connector is placed as close to the patient's airway as possible so that there is no unnecessary dead space between the patient and the monitoring site; then the $ETCO_2$ measured will closely reflect alveolar CO_2. When a condenser, humidifier, or filter is used in the circuit, the airway connector is ideally placed between the tracheal tube and the heat and moisture exchanger or filter, despite the increased risk of water getting into the sampling tube.

Although the potential for a wide range of gas flows per minute is available, gas flow of 150 ml/min is considered optimal as the capnographs generated are reliable in both children and adults. Sinusoid deformities are produced in the capnograph curves when higher gas flows are used for monitoring very small patients.

Response time

A disadvantage of the sidestream analyser is that analysis is delayed slightly because gas is routed through a catheter to the capnometer. The transit time, which is the time taken for the sample to be transferred from the sampling port along the catheter to the analyser, depends on the length and diameter of the catheter, the rate at which the gas is aspirated, and the viscosity of the sample. The transit time delay can be calculated from the formula

$$\text{delay} = \pi D^2 L / (4Q)$$

where D = diameter of the sampling tube, L = length of the sampling tube, and Q = sampling flow rate. The transit time of the sidestream analyser is normally not important because the signal is delayed for less than a second and is not seriously distorted. The transit time with long sampling tubes such as are used with the shared mass spectrometer can be as long as 20 seconds—this has the disadvantage that the capnogram being displayed lags considerably behind the one being measured. Consequently, the shorter the length of the sampling tube the faster the transit time and the greater the accuracy, and the more representative the displayed capnogram is of current CO_2 changes.

As in the mainstream analysers, the inherent rise time (also referred to as the response time) of sidestream capnometers causes a slight additional delay. This rise time, defined as the time taken by the capnographic cell to register from 10% to 90% of a step change in concentration after the gas has entered the analysis cell, depends on the size of the sample chamber and the gas flow. Rise time is a key determinant of a capnograph's ability to accurately reflect changes in CO_2 at the airway. Small analysing chambers of about 1 ml volume with an appropriate flow can have rise times as short as 0.1 second, and most modern instruments have 95% rise time of about this duration. The efficiency of the electronics and signal processing software also determines the duration of the rise time. Total delay time is the rise time plus the transit time and must be less than the respiratory cycle time (time taken for one breath) or the analyser will give falsely low end tidal CO_2 and high baseline CO_2 readings.

Gas flow rates

Changes in the gas sampling flow also affect the shape of the capnographic curves. The aspirated gas sample can mix with other gas in the tube if slow rates of flow and long catheters (> 2 m) with large lumen are used and will give unacceptably long rise times. This mixing can produce a capnogram where the curves are raised above the baseline and are sinusoidal in form without a plateau. Placing the airway connector as close as

31

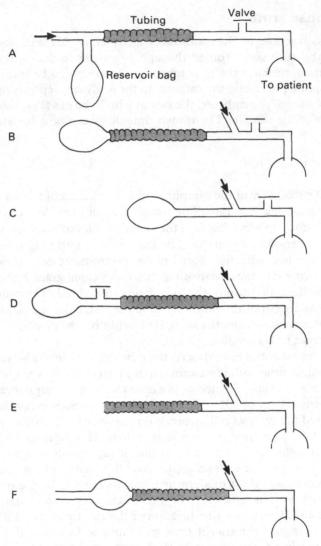

Fig 3.7 Mapleson classification (A–F) of semiclosed rebreathing anaesthetic systems. Arrow indicates entry of fresh gas to system

possible to the tracheal tube reduces the potential for fresh gas from the anaesthetic machine mixing with expired gas, especially when a partial rebreathing system (Mapleston D and Ayre's T piece) is used—this would produce an erroneous low value for the end tidal sample because of the dilution. The Bain is the commonly used version of the Mapleson D, and like the Lack (Mapleson A) is an example of a coaxial system (fig 3.7, 3.8).

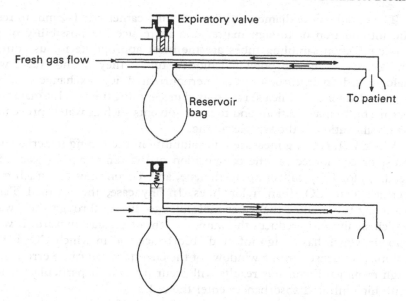

Fig 3.8 Coaxial anaesthetic breathing systems: (top) Bain system (Mapleson D); (bottom) Lack system (Mapleson A)

High aspirating flow rates can also give incorrect results: they can cause a considerable pressure drop across the sampling line, and the output of the analyser depends on pressure. Restricting the length of the sampling tube may be a disadvantage when the anaesthetic machine needs to be some distance from the patient, as in head and neck surgery. From a practical point of view, however, the sidestream analyser system is easier to use in patients breathing spontaneously who are not anaesthetised than is the mainstream analyser.

Tips, tubes, and traps

In an ideal situation, the tip of the sampling catheter would be positioned just above the carina to obtain an accurate sample of alveolar gas. This is impractical, however, as the risk of aspirating secretions and water into the apparatus is considerable, and there is potential for irritation of the airway. The tip is routinely sited at the proximal end of the tracheal tube by using a T piece adapter, analogous to the positioning of the infrared measuring cell in mainstream analysers. As in mainstream analysers, it is important that the adapter is placed next to the tracheal tube distal to the catheter to avoid introducing extra dead space, which would give falsely high CO_2 readings. A syringe needle should not be inserted through catheter mounts to achieve access to the respired gas because of the risk of needle stick injury.

The optimal bore diameter of the sampling catheter is 1–2 mm to resist the introduction of foreign matter and to reduce the possibility of gas mixing. Teflon sampling tubes are the most appropriate for use during anaesthesia as, unlike polyvinylchloride tubes, they do not react with halogenated hydrocarbons. The perflourinated ion exchange polymer Nafion (Dupont Chemicals) is also used in sampling tubes. This material is permeable to many cations and to polar solvents such as water, preventing its accumulation in the sample tubing.

When CO_2 is being measured its solubility in the tubing materials used must be considered, as the composition of the sampling line can affect readings for CO_2. Silicon, polyethylene, and Teflon lines are much more permeable to CO_2 than nylon lines. In any case, the warmed Teflon sampling tubes allow the water molecules to diffuse through their walls, which significantly reduces the water content of the gas mixture. If water vapour, which has a high infrared absorbance and in which CO_2 is very soluble, condenses on the windows of the cuvette it can cause erroneously high readings. Erroneous results will occur if water or particulate matter with high infrared absorbances enter the cell.

Hydrophobic sampling tubes resist the entry of water into the analysis cells. As well, inserting water traps between the distal end of the tubing and the analyser can reduce this problem. Water condensing in the tube is collected and absorbed in the water trap, which is designed to allow minimal turbulence to gas flow. It relies on gravitational forces to separate drops of water from the gas stream. These traps need to be dried out often and water prevented from accumulating in them. When Teflon tubes are used, a water trap is not required, but blockage of the tubing by mucus plugs, difficulties in cleaning, and the higher cost are disadvantages. The temperature of the patient is usually about 37°C, while that of the instrument cell is 25°C, causing a pressure difference which results in an overestimate of PCO_2 by 0.15% (0.15 kPa).

Exhaust gases

The aspirated gas sample in sidestream analysers often contains anaesthetic gases and so the exhaust gas from the capnometer should not be vented into the room but routed to a gas scavenger or returned to the breathing system to eliminate anaesthetic gas pollution in theatre. Returning the gas to the patient leaves open the possibility of transmitting infection from previous patients who have exhaled into the conduits. Disposable breathing systems or sterilisation between patients would overcome this hazard, but at considerable expense. Active scavenging of the aspirated gas is more practical, except in closed circuit anaesthesia. Capnometers should be calibrated with the scavenging system incorporated because the scavenging system may resist gas flow and pressure in the

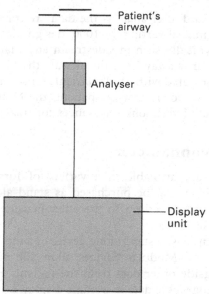

Fig 3.9 The analyser of the proximal-diverting N-1000 is located midway between the patient and the display unit, avoiding many of the problems associated with sidestream or mainstream capnographs

analysing chamber may be affected, or the scavenger may generate a vacuum, affecting the sampling flow of the capnometer and producing errors in the resulting capnograph.

In closed circuit anaesthesia systems, sidestream analysers aspirate about 200 ml/min of oxygen, which is about as much as the basal requirement for the patient. With ultra low flows, instead of scavenging the aspirated gas, in order to conserve gas flow it is returned to the breathing system through a bacteria filter in the expiratory limb of the patient circuit. Alternatively, compensation may be made for this gas consumption in the capnometer by providing extra equivalent fresh gas inflow when a closed system is in use. For the same reasons, excessive sampling rates should also be avoided when paediatric breathing systems are used. Manufacturers produce adapters to cope with the problems of capnographic monitoring in small children.

Proximal-diverting capnometry

A new gas sampling system claims to combine the desirable features of the traditional sidestream and mainstream design capnographs. In the Nellcor N-1000 multifunction monitor a proximal-diverting gas sampling system removes the sensor from the airway but keeps it near the patient— that is, it is located midway between the patient and the display unit (fig 3.9). Gas for analysis is transported to the sensor, which is located in the

35

patient module, and after analysis the data are transmitted to the display unit. The proximal-diverting N-1000 design deals with many of the disadvantages typically seen in sidestream and mainstream capnographs: because the sensor is away from the airway, the N-1000 does not present the problems associated with a bulky, fragile sensor mounted on the airway; and because the sensor is near the patient, the N-1000 does not have the problems associated with longer distances for transport of samples.

Selecting a capnometer

Capnometers are available in a variety of formats to suit different requirements. They can be purchased as stand alone units, as modular units that interface directly with the other bedside monitors, or as combinations of these forms.

Whether to purchase a stand alone device or a modular unit depends on a number of factors. Modular devices allow the capnograph data to be integrated alongside other data from the patient. Having haemodynamic and electrocardiographic data in a single display is useful in evaluating the entire clinical picture as well as simultaneously observing trends in several variables. Because separate digital and oscilloscopic displays are not necessary for every variable measured, the modular monitors are usually less expensive than stand alone units. Stand alone devices are useful when portability and size are important—for example, during transfer of patients between hospitals or between units in a hospital.

Calibration

Capnometers for monitoring during surgical operations must be able to detect carbon dioxide in the range of 0–10% (up to 10 kPa (76 mm Hg)), except in rare circumstances such as malignant hyperpyrexia or gross hypoventilation, where an extended range would be useful. An accuracy of 10% is probably sufficient because in clinical practice the capnographic trends are more informative than actual readings. Most CO_2 monitors allow the user to select either kPa or percentage as the unit of measurement. Changes in ambient pressure are automatically compensated.

To reduce the possibility of erroneous CO_2 readings due to faults in the monitoring system, the sampling line, airway adapter, monitor, and mains cable should be checked visually before use. The sampling line and airway adapter are usually disposable and should be carefully inspected for cracks or other potential causes of leaks. Water trap containers should be emptied before monitoring begins.

Various techniques that produce signals can confirm the operation and accuracy of the capnometer. The monitor can be roughly calibrated by the clinician breathing normally into the sampling tube. A mean reading of

$ETCO_2$ of 5.3 kPa (40 mm Hg) or approximately 4.5–5.5% would be expected. The $ETCO_2$ is usually displayed as partial pressure by the atmospheric pressure. A more accurate technique by which to confirm the accuracy of end tidal values uses canisters of gas mixtures containing known amounts of CO_2 (5%, for example) in a mixture of anaesthetic gases and vapours; regular checks using this easy method are recommended.

Calibration routines should use the same type of sampling catheter as will be used when the analyser is monitoring patients. If the normal 2 m long sampling catheter is not used during calibration, errors in end tidal carbon dioxide concentrations may result during subsequent clinical use due to the absence of a significant pressure drop across the ends of the tube.

Some of the more sophisticated modern instruments incorporate means for electronic calibration with optical filters, beam occluders, or a sample cell containing carbon dioxide. These machines automatically carry out a zero calibration to ensure that the baseline measurement is a true zero measurement of CO_2; they also do a span calibration, correcting the measurement scale to ensure that the instrument is accurate over its performance range. They have the capability to detect changes of barometric pressure and automatically correct the CO_2 reading.

Automatic calibration can be a disadvantage in that it requires time and may interrupt monitoring during a critical period of patient care. Complete calibration tests should be required infrequently, and rapid tests no more than once a day. Equipment drift is caused mainly by accumulation of secretions in the light path of the analyser. Nellcor's Ultracap mainstream analyser is calibrated at the factory and requires no operator calibration at start up or during prolonged monitoring.

Alarms

Capnographs will sound an alarm during apnoea when another breath fails to arrive. The time from apnoea to alarm is often adjustable and should be set for no greater than 15 seconds. The data display of the capnograph needs to be visible from a reasonable distance. Awareness of the potential for artefactual changes in the capnogram will avoid their misinterpretation as physiological events.

Mass spectrometry

Early this century Thompson and Aston developed mass spectrometry, based on the concept of separating charged particles (ions) by their mass. The technique was initially practically applied in isotope research, but in the past two decades mass spectrometry has been more extensively used in

37

the United States to measure several gases simultaneously. The technique's earliest role in medicine was in investigative respiratory physiology and pulmonary function testing; it was not used in anaesthetic monitoring until the early 1980s. Two types of mass spectrometer have become important medically: the magnetic sector, with fixed detectors, and the quadrupole. They have been used either as a dedicated instrument providing continuous monitoring of single patients or as a shared instrument (multiplexed) providing discontinuous monitoring of several patients in sequence.

With its capability of measuring all respiratory and anaesthetic gases, mass spectrometry remains the most versatile method of analysing exhaled gases. In contrast, infrared capnometry is limited to analysing molecular compounds containing more than two elements.

The ionisation process

All clinical spectrometers behave as sidestream analysers. In both the magnetic sector and the quadrupole types of mass spectrometer the gas to be analysed is aspirated in very small quantities into a high vacuum chamber with a very low pressure (10^{-5} mm Hg). An electronic beam ionises a proportion of the gas molecules, stripping the individual molecules of one or more electrons and transforming them into positively charged ions, which are then accelerated by an electric field into the final analysing chamber of the spectrometer.

In the magnetic sector analyser, a powerful magnetic field in the final chamber perpendicular to the ion trajectory deflects these fast moving ions into circular paths whose radii of curvature are proportional to the charge to mass ratio of the ions. Lighter ions tend to be deflected more, producing an arc with a smaller radius (fig 3.10). For each gas to be measured, a detection plate must be placed at a specific location so that the ions unique to that substance will strike it. Analysis of the spectrum of ions depends on the ions per minute registered at the detection plates, which can then be converted into an electric current. This current can be interpreted by a microprocessor incorporated in the mass spectrometer to display waveforms, trends, or numerical values of the gases' concentration in the sample.

The more popular quadrupole type of analyser has a mass filter, which depends on a combination of direct current and radio frequency potentials applied to an arrangement of four parallel rods (the quadrupole). After ionisation the gas molecules are focused into a narrow ion beam and are accelerated by an electric field towards the detector, passing through the quadrupole. Rapid controlled changes in the quadrupole's electrostatic field force ions of a fixed charge to mass ratio to pass through selectors into a detector. The rate at which ions of a single charge to mass ratio impinge

38

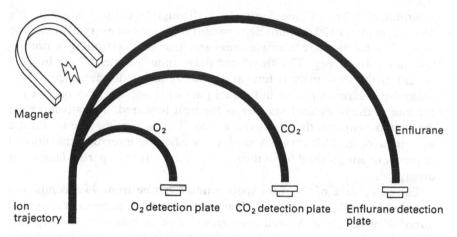

Fig 3.10 The magnetic sector analyser

on the detector is proportional to the concentration of that particular gas in the original mixture. Varying the ratio of the direct current and radio frequency potentials makes it possible to scan the complete spectrum in a few milliseconds. Sampling the detector output at appropriate times in the scan allows the spectrum analyser to generate a plot of the concentration of each gas.

The ionisation process in the mass spectrometer generates mainly singly charged ions or ions missing a single electron. As mass spectrometry measures individually different concentrations of gases based on the charge to mass ratio, the most important variable is the molecular weight. As happens with overlapping absorption bands in infrared spectrometry, different molecules of the same molecular weight (for example, CO_2 and N_2O) are therefore not distinguishable by a mass spectrometer on this basis. The ionisation process in the mass spectrometer also fragments the components of the gas sample to be analysed, and overcomes this problem by incorporating empirical algorithms to distinguish particular pairs of molecules by analysing these molecular fragments.

Placement and monitoring

Mass spectrometers have traditionally been too bulky to use at the bedside. They are expensive instruments and are mainly used in the United States as a shared instrument providing discontinuous monitoring of several patients in several operating theatres or of all patients in an intensive care unit. A single mass spectrometer is placed at a remote convenient location and gas samples are entrained from up to 31 patients'

39

breathing circuits on a time sharing basis through long capillary tubing at a flow rate of about 150 ml/min. Each patient is monitored every 3 minutes or so if 16 patients share a single mass spectrometer, assuming a normal respiratory frequency. The significant delay time and the sharing by each patient of the total analysis time available are major disadvantages. Continuous measurement in an individual patient is possible, but only at the expense of the other analysis sites as the unit is shared. The advantage of the shared system is that it is economical. The facility may be useful if a user in a central observation position wishes to interrupt the normal sequence of analysis and focus the mass spectrometer to a particular patient briefly.

The delay time of the mass spectrometer can be from 45 seconds to 5 minutes depending on factors such as the transit time, number of sites, and duration of sampling. Shared mass spectrometers may have two separate delay times because the rates of sampling flow and measuring flow are not similar. Gas sampling flow is adjusted to a rate sufficient to enable a 20 second gas sample to be stored in the sampling line. During gas measurement, flow rate is doubled and the 20 second sample of gas is analysed within about 7 seconds, after which the display is updated. The 7 seconds is sometimes called dwell time and is the actual time of delay—that is, when monitoring is continuous, the gas sample arrives at the analyser 7 seconds after being sampled.

Inspiratory and expiratory concentrations of oxygen, nitrogen, carbon dioxide, and anaesthetic vapours are the variables usually displayed. The addition of a known flow rate of an inert gas—for example, argon—in small quantities enables respiratory volumes and metabolic rate to be calculated simultaneously.

Another disadvantage is that if the spectrometer malfunctions no patients are monitored because it is usually used on a shared basis. In contrast, if an infrared analyser breaks down the other units will continue to monitor in the other theatres or intensive care beds. Manufacturers have solved this problem by incorporating sidestream infrared analysers into the mass spectrometer system. A dedicated mass spectrometer can of course be used to monitor a single patient continuously, but this is usually too costly.

Photoacoustic spectroscopy

The Danish technique of photoacoustic spectroscopy is based on the same principles as conventional infrared gas analysers—namely, the ability of carbon dioxide, nitrous oxide, and anaesthetic agents to absorb infrared light. The techniques differ in the detection techniques applied: acoustic versus optical methods. In photoacoustic spectroscopy, pulses of infrared light are applied to a gas mixture. The frequency of these pulses lies within the audible range, and the frequency of the infrared light can be tuned across the infrared spectrum. When the frequency of the infrared light matches the optimum absorption frequency of a gas in the sample, the gas is caused to expand in a pulsatile fashion matching the audiofrequency of the pulses in the infrared beam. These audiofrequency expansions are detected by a microphone with a sound output proportional to the concentration of gas in the sample.

The advantages claimed for photoacoustic spectroscopy over conventional systems are that its equipment is simpler, smaller, and has high accuracy. In a standard infrared analyser the energy absorbed by the gas sample is measured indirectly by measuring the transmission through the measurement chamber and comparing it to that through a reference cell. With photoacoustic spectroscopy the amount of infrared light absorbed is measured directly by measuring the sound energy emitted; thus, photoacoustic measurement is more linear and less inclined to drift. This means that photoacoustic spectroscopy is highly accurate, with very little instability. Zero point drift is almost non-existent as zero is always reached when there is no gas present; if no gas is present no acoustic signal is registered. Another difference is that infrared analysers are optical: the transmitted signal is picked up by a photosensor. In the photoacoustic system the microphone, which is used as a detector, confers a high degree of stability and reliability on the analytical procedure, and as a result the monitor rarely requires calibration. The photoacoustic system therefore needs less preventive maintenance.

One example of an instrument developed for photoacoustic gas measurement is the Bruel & Kjar anaesthetic gas monitor 1304[1] (fig 3.11). The gas sample is drawn into the measurement chamber through a flow regulator. This makes the sample flow rate independent of changes in the patient's airway pressure. Light from an infrared source is reflected off a mirror towards a window in the measurement chamber. Before the light enters the chamber it passes through a spinning chopper wheel, which causes it to pulsate. To allow differentiation between the three signals caused by carbon dioxide, nitrous oxide, and anaesthetic agent, the chopper, which consists of three concentric bands of holes, divides the incident infrared beam into three different parts which pulsate at three different frequencies.

The divided light beam then passes through a specially constructed optical filter which consists of three separate and different filters. The optical filters are mounted in front of the measuring chamber, and each corresponds to one of the parts of the light beam. Each filter allows only light of a specific wavelength to pass through; the frequencies and wavelengths of the incident light are thus optimised to match the infrared absorption spectrum of the three gases to be measured.

In the measurement chamber the three different beams, which now differ in wavelength and pulsation frequency, each excite one of the gases. Figure 3.12 shows the infrared absorption spectra for the three gases. The

41

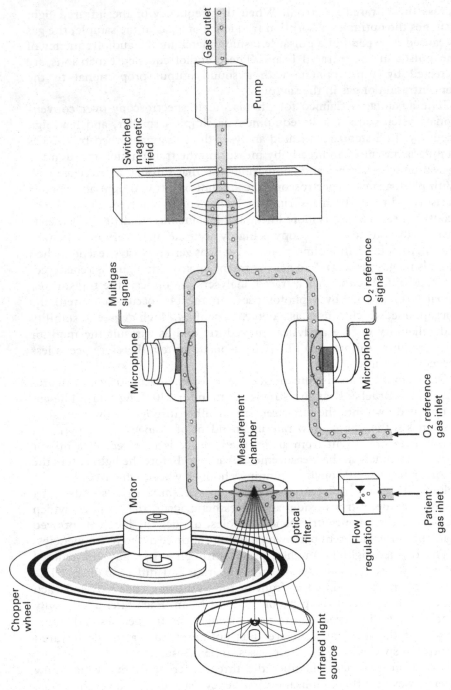

Fig 3.11 The Bruel & Kjaer acoustic measurement anaesthetic gas system 1304. (Reproduced with kind permission of Bruel & Kjaer)

Gas outlet

Pump

Switched magnetic field

Multigas signal

O₂ reference signal

Microphone

Microphone

O₂ reference gas inlet

Measurement chamber

Motor

Optical filter

Flow regulation

Patient gas inlet

Chopper wheel

Infrared light source

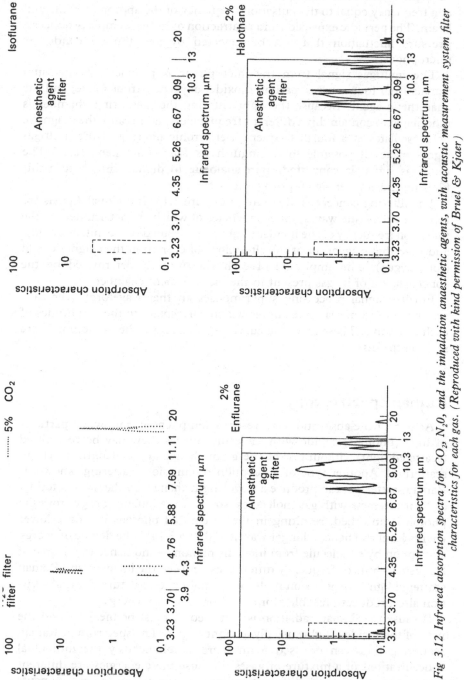

Fig 3.12 Infrared absorption spectra for CO_2, N_2O, and the inhalation anaesthetic agents, with acoustic measurement system filter characteristics for each gas. (Reproduced with kind permission of Bruel & Kjaer)

absorption of the incident light will cause each gas to expand and contract at a frequency equal to the pulsation frequency of the appropriate infrared beam. The periodic expansion and contraction of the gas sample produces a pressure fluctuation that can be detected by a optimised condenser microphone.

The multigas signal from the microphone is produced by pressure signals from four sources: carbon dioxide, oxygen, nitrous oxide, and the anaesthetic agent. Because the light entering the measuring chamber is modulated to contain three different frequencies, and because the magnetic field oscillates at a fourth frequency, electronic filtering of the multigas signal makes it possible to distinguish its four component parts. The multigas signal is converted from analogue to digital form before this filtering and further signal processing occur.

The filtering concept is illustrated in figure 3.13. Individual signals are represented as sine waves, the frequencies of which are determined by the pulsation frequency of the incident light (for carbon dioxide, nitrous oxide, and anaesthetic agent) or of the alterations of current in the magnetic field (for oxygen). The amplitude of each sine wave is determined by the concentration of its gas present in the measurement chamber.

Further signal processing superimposes another waveform over each sine wave to form real time curves which correspond to the amplitudes of each sine wave. These real time curves—for example, the capnogram—are seen on the display.

Raman spectroscopy

As part of the absorption and re-emission phenomenon, when particles of light (photons) collide with molecules of gas they may be re-emitted (scattered light) without loss of energy or change in wavelength (Rayleigh scattering). Another type of absorption re-emission scattering, known as Raman scattering, can produce a wavelength change. In Raman scattering, as light interacts with gas molecules, some of the kinetic energy from the photon is absorbed, resulting in the re-emitted photons having a lower energy level and hence a longer wavelength (fig 3.14). The degree of energy absorption by a molecule from incident photons—and hence the degree of frequency shifts or frequency differences between the incident and Raman scattered light—depends on molecular weight and structure. This absorption also produces unstable vibrational or rotational energy states.

Because the Raman radiation is scattered, it must be measured off the path of the incident light, usually at right angles. The spectrum of Raman scattering lines can be used to measure simultaneously the individual concentrations of a mixture of gases. Because these energies are different for specific bonds and different molecules, the frequency components

44

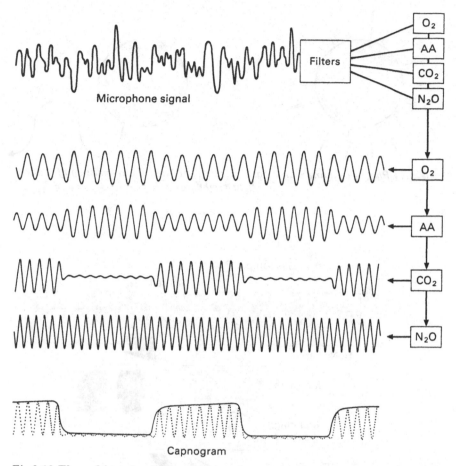

Fig 3.13 *The multigas signal from the microphone is electronically filtered into its four component parts. The amplitude of each generated sine wave is determined by the concentration of the appropriate gas present in the measurement chamber. Further signal processing superimposes another waveform over each sine wave to form real time curves which are seen on the display. (Reproduced with kind permission of Bruel & Kjaer)*

present in Raman scattered light provide specific chemical identification of the molecules irradiated.

The Raman radiation (Raman light) is of low intensity, since only 0.1% of the incident light is scattered and only 0.1% of the scattered light experiences a Raman shift. The gas sample is aspirated into an analysing chamber, where the sample is illuminated by a high intensity light source, usually an air cooled monochromatic argon laser with a wavelength of

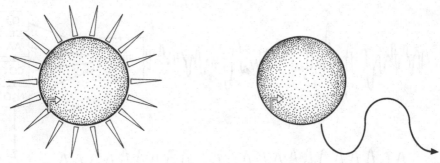

Fig 3.14 *Photon colliding with a gas molecule (left) and the re-emission of scattered light (right)*

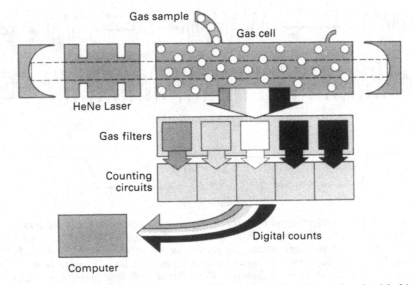

Fig 3.15 *The basis of Raman scattering for gas analysis. (Reproduced with kind permission of Bruel & Kjaer)*

488 nm (fig 3.15). The scattered light is then passed through a spectrometer. Sensitive photomultiplier tubes have replaced long exposures of fast photographic emulsions for detection of the peaks.

The concept of Raman scattering was discovered by CV Raman and KS Krishnan in 1928, but its use in medical monitoring is only recent. Prototype instruments for continuous CO_2 measurement are now becoming available, and the early results look encouraging. Raman scattering has

46

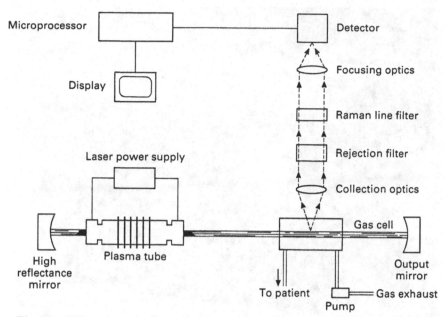

Fig 3.16 The Raman spectrometer. (Reproduced with kind permission of D Westen-skow, Anesthesiology 1989;70:350–5)

been incorporated into the newer anaesthetic monitors (Rascal monitors; Albion Instruments, Salt Lake City, Utah) to identify and quantify instantly CO_2, inspired and expired oxygen, nitrogen, and inhalational anaesthetic agents. Figure 3.16 shows the components of the first commercially available Rascal. A sample cell (1.0 ml) is positioned between the laser plasma tube and the output mirror of a 40 mW aircooled laser. The argon laser beam passes directly through the gas cell. Gas molecules in the sample cell emit Raman scattered light with characteristic wavelengths. Lenses collect the scattered light from the sample cell and focus it through interference filters, which select light that corresponds to each of the gases to be measured. Photomultiplier tubes convert the scattered light to photon counts. The number of counts received during 100 ms is used to determine the concentration of each gas. The digital display shows the average plateau values for inspired and expired concentration of O_2, N_2, CO_2, N_2O, halothane, isoflurane, and enflurane and the peak expired CO_2 concentration.[2]

Ohmeda's Rascal II monitor (Ohmeda, Hatfield, Hertfordshire; fig 3.17) contains a precisely regulated laser beam which passes through a gas cell containing a sample of gas from the patient's breathing circuit.[3] When a photon of light from the laser beam collides with a gas molecule in the cell,

Fig 3.17 The Ohmeda Rascal II anaesthetic gas monitor

energy excites the vibrational and rotational modes of the molecule. As the molecule loses energy after the collision, it re-emits scattered light at a lower energy and consequently at a greater wavelength. Photodetectors identify and measure each gas. Because wavelength "shifts" are different for each gas, Raman scattered light allows precise identification of gases (fig 3.18). The laser-Raman sensor uses proprietary filters and photon counting circuits to digitally measure the scattered photon flux at multiple wavelengths. These digital values are then converted to precise gas concentrations.

Figure 3.18 shows the major Raman peaks present in a sample of gas containing N_2, O_2, and CO_2 and a trace of water vapor. The amplitudes of the peaks are proportional to the concentrations of the individual gases in the sample; therefore, the intensity of Raman scattered light at specific frequencies gives an absolute measure of gas concentration. Because the peaks are clearly separated, gas concentrations are measured independently.

Advantages

Monitors based on Raman scattering have advantages over other systems (box). The ability of monitors based on Raman scattering to precisely

48

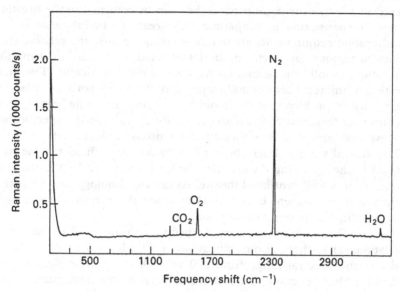

Fig 3.18 The Raman scattering spectrum. Sample volume contained 5% carbon dioxide, 12% oxygen, 83% nitrogen, and a trace of water vapour. (Reproduced with kind permission of D Westenskow, Anesthesiology *1989;70:350–5)*

Advantages of monitors based on Raman scattering

Over multiplexed mass spectrometry
- Simple calibration
- Fast response time
- Independent analysis of each gas

Over infrared analysis
- Reponse time
- Differentiation among volatile anaesthetic agents
- Detection of nitrogen and oxygen

Over mass spectrometry
- Dedicated to one patient

identify anaesthetic agents helps to reduce the risk of incorrect mixtures or dosages of vapor agents, and the ability to identify nitrogen facilitates early detection of leaks or disconnections of the anaesthetic circuit.

Measurement techniques based on Raman scattering are extremely sensitive and have been used for atmospheric monitoring, where concentrations are measured in parts per million. Since the sampled gases are not altered by the measurement process they can be returned to the breathing system. Response time is compatible with breath by breath analysis.

Calibration requirements are not demanding because the relative sensitivities to various gases remain absolutely fixed. After an initial multigas calibration, monthly recalibration with room air is sufficient. The incorporation of microprocessors and empirically derived algorithms has overcome earlier problems with resolution during multiple gas analysis. Performance factors such as accuracy, rise time, and stability specifications of these monitors have been found to encompass clinical needs.

The Rascal's argon laser cooling fan generates sufficient noise to be noticed in the operating theatre, but the heat dissipated by the fan is not noticeable in a well ventilated theatre. As laser technology continues to be refined, a more efficient laser should alleviate the concerns about noise, heat, weight, and power consumption.

Considerable overlap of Raman spectra can occur for the volatile inhalational anaesthetic agents, which have multiple, closely spaced Raman peaks in the spectral range from 200 to 1500 cm^{-1}. The Raman peaks picked for filter selection have maximum intensity and minimum overlap, yet the resulting limited filter bandwidth still permits substance throughput from adjacent Raman lines arising from other gases. Linear matrix analysis of signals from filters for the three anaesthetic agents (isoflurane, enflurane, and halothane) is used to determine the identities and relative concentrations of the anaesthetic agents. A peak common to all three agents is used to determine the absolute concentration of the agent. The signals from all four filters are analysed simultaneously, and the anaesthetics are individually identified and their concentrations measured independently.

Summarising methods for measuring carbon dioxide continuously in anaesthetic practice: infrared analysis is preferred to mass spectrometry. Of the two types of analyser systems available, the sidestream system is often preferred because of the expense involved in the repair of accidental damage sustained to the delicate components in the mainstream adaptor. The sidestream analyser system, although not as accurate as the mainstream analysers, is easier to use in patients breathing spontaneously who are not anaesthetised than is the mainstream analyser. The latest technique of Raman scattering has the potential to combine the lower expense of infrared with the multigas capability of mass spectrometry. The photoacoustic monitor is compact, easy to use, highly accurate and reliable, and needs calibration less frequently.

Transcutaneous CO_2 monitoring

If areas of skin are warmed to 44°C dissolved gases such as CO_2 diffuse through the skin and out of the body. This principle is the basis of transcutaneous PCO_2 measuring devices. Transcutaneous CO_2 monitoring was introduced in the early 1980s following the increase in popularity of transcutaneous pO_2 monitoring, which was useful in preventing hyperoxia in neonatal care. Pulse oximetry has mostly replaced transcutaneous pO_2 monitoring, and the transcutaneous CO_2 approach has not established a definite niche.

Transcutaneous CO_2 devices usually consist of a small sensor, a heater, and a temperature sensing thermistor; they are attached with adhesive to the skin to be airtight. The sensor is composed of a pH sensing glass electrode developed by Severinghaus in 1958, similar to the pH electrodes in blood gas analysis machines. The carbon dioxide that is released through the skin diffuses through a $25\,\mu m$ permeable Teflon membrane and combines with water, resulting in the production of H_2CO_3, which then dissociates into H^+ and HCO_3^-. The pH sensor in the transcutaneous CO_2 electrode measures the concentration of hydrogen ions in a solution, the change in which is proportional to the change in PCO_2. The results are referenced to a silver or silver chloride electrode and calibrated to the CO_2 tension. The heater warms the skin locally and the thermistor regulates the skin temperature underneath the probe.

Haemodynamically stable adult patients with uncompromised skin perfusion show a significant correlation between transcutaneous and arterial PCO_2. Since the transcutaneous PCO_2 system measures CO_2 in the tissues, normal transcutaneous values are higher than arterial values; however, the difference is relatively constant. The normal transcutaneous value for an adult is about $8\,kPa$ $(60\,mm\,Hg)$ when the $PaCO_2$ is $5.3\,kPa$ $(40\,mm\,Hg)$. Under stable circumstances, a change in transcutaneous PCO_2 will mirror a change in $PaCO_2$ but is delayed 3–4 minutes. Local metabolic production of carbon dioxide, higher probe temperature, and capillary blood flow are among the factors which contribute to the difference between transcutaneous and arterial CO_2, which is about $2.6\,kPa$ $(20\,mm\,Hg)$ in neonates and $3.3\,kPa$ $(25\,mm\,Hg)$ in adults if the probes are calibrated and used at the same temperature.

Manufacturers have incorporated correction factors into devices so that the transcutaneous CO_2 values displayed are close estimates of the $PaCO_2$. The transcutaneous value is usually "corrected" by dividing it by a number near 1.5. When these corrections are incorporated, the displayed value is normally within 10% of the arterial CO_2 value. As the transcutaneous PCO_2 reflects tissue CO_2 tension more than arterial tension, there is no other available method of calibration or validation, whereas oximetry readings of oxygen saturation can be confirmed by arterial blood sampling and end tidal CO_2 values can be checked against arterial CO_2 tensions. Reapplying and recalibrating the instrument to the same patient can help to validate transcutaneous readings, but this is time consuming. These limitations emphasise the need to study transcutaneous PCO_2 trends rather than interpreting isolated numerical values.

Limits on transcutaneous monitoring

- Relatively slow response and calibration times limit use in acute situations such as cardiac arrest and cardiorespiratory resuscitation
- Absence of established physiological norms or critical levels limits use to observations of trends where patients act as own controls
- Period required to establish data for trend analysis limits role to non-acute clinical situations
- New probes need up to an hour to warm up so that electrode solution is evenly distributed

---

The relation between transcutaneous and arterial PCO_2 in the presence of pulmonary disease or in the haemodynamically unstable patient is less clear. Studies have shown that in the presence of severe hypotension transcutaneous PCO_2 rises disproportionately to arterial PCO_2.

Most transcutaneous CO_2 monitoring devices combine a Clark-type O_2 electrode and a Severinghaus-type CO_2 electrode in a single unit. They are easy to apply without skin preparation. One of the major disadvantages of transcutaneous monitoring is the relatively lengthy response time (3–15 minutes), which is influenced considerably by probe temperature. The higher the probe temperature, the faster the observed response time. Therefore, a high skin temperature is necessary to obtain clinically useful transcutaneous PCO_2 values with the least possible delay. Warming the skin increases gas permeability across the cutaneous barrier by softening the keratin layer. But warming also increases local production of CO_2 and requires a site change every 4 hours to avoid burns. The thermistor should locally regulate the skin temperature under the probe, but thermostat failure could cause burns. In conditions of reduced tissue perfusion secondary to peripheral vasoconstriction or thick (adult) skin the measurement may be erroneously high. Most research work in transcutaneous monitoring has been carried out in subjects with thin skin (neonates), and studies have shown a high degree of accuracy with less variability than in adults.

In the critical care setting, information from transcutaneous monitoring must be considered as only one monitoring contribution, of unclear significance, in the overall appraisal of a particular clinical situation. Changes in vital signs and other variables are likely to be evident minutes earlier than changes in transcutaneous PCO_2 in any case.

The role of transcutaneous monitoring in adult critical care is as yet to be defined. Due to the long response and lag times and the less than ideal relation to arterial values, it is hard to imagine transcutaneous monitoring establishing itself as a frontline monitor in rapidly changing clinical situations that frequently occur in intensive care. Although they are non-

Fig 3.19 The EASY CAP disposable end tidal carbon dioxide detector

invasive, accurate, and relatively inexpensive, transcutaneous monitors offer little extra information to that already provided by other monitors used in intensive care.

The end tidal carbon dioxide detector

A disposable end tidal carbon dioxide detector (EASY CAP, Nellcor, Hayward, California) has recently been introduced; in it a pH sensitive indicator changes colour when exposed to carbon dioxide (fig 3.19). The colour of the end tidal detector varies between expiration and inspiration as carbon dioxide concentrations increase and decrease. The colour changes from purple (when exposed to room air) to yellow (when exposed to 4% carbon dioxide).

Housed in a plastic container, the detector has an internal volume of 38 ml and a resistance to flow of less than 0.3 kPa at 60 l/min flow; it weighs less than 30 g and has standard (15 mm) connection ports. This compact detector is connected to the proximal end of the placed tracheal tube distal to the catheter mount in the anaesthetic system. The device responds sufficiently quickly to detect changes of carbon dioxide breath by breath. The method is simple and takes only seconds to perform.

One of the limitations of the device is that carbonated beverages or sodium bicarbonate in the stomach produce significant CO_2 waveforms on capnography and colour changes on the end tidal carbon dioxide detector for at least six breaths after oesophageal intubation. Consequently, this interval should elapse before the tracheal tube position is decided.

Colour changes must also be interpreted with care during cardiopulmonary resuscitation. Pulmonary perfusion is significantly decreased or absent during cardiac arrest, and the end tidal carbon dioxide detector may fail to register a colour change after intubation even though the tube has been correctly inserted in the treachea.

Overall, the end tidal carbon dioxide detector offers an inexpensive and reliable alternative for the qualitative estimation of CO_2 in situations where capnometry is not readily available.

1 McPeak H, Palayiwa E, Robinson G, Sykes M. An evaluation of the Bruel and Kjaer monitor 1304. *Anaesthesia* 1992;47:41–7.
2 Westenskow D, Smith K, Coleman D, Gregonis D, van Wagenen R. Clinical evaluation of a Raman scattering multiple gas analyser for the operating room. *Anesthesiology* 1989;70: 350–5.
3 Lockwood GG, Landon MJ, Chakrabarti MK, Whitwam JG. The Ohmeda Rascal II. A new gas analyser for anaesthetic use. *Anaesthesia* 1994;44:44–53.

Further reading

Eichhorn J. Monitoring standards for clinical practice. *International Anesthesia Research Society Review Course Lectures* 1988;1:113–9.
Gravenstein JS, Paulus D, Hayes T. *Capnography in clinical practice*. London: Butterworth, 1989.
Kalenda Z. *Mastering infrared capnography*. Utrecht: Kerckebosch-Zeist, 1989.
Paloheimo M, Valli M, Ahjopalo H. *A guide to CO_2 monitoring*. Helsinki: Datex Instrumentarium, 1988.
Raemer D, Chalang I. Accuracy of end-tidal carbon dioxide tension analyzers. *J Clin Monit* 1991;7:195–201.
Raemer DB, Philip JH. Monitoring anesthetic and respiratory gases. In: Blitt CD, ed. *Monitoring in anesthesia and critical care medicine*. 2nd ed. New York: Churchill Livingstone, 1990:373–86.
Sykes M, Vickers M, Hull C. *Principles of measurement and monitoring in anaesthesia and intensive care*. Oxford: Blackwell Scientific, 1991.

4 Normal and abnormal waveforms

As explained in chapter 1, the normal CO_2 waveform (the capnogram) reflects the different stages in breathing. As the shape of the capnogram is virtually identical in all healthy patients, any alteration in shape from normal must be recognised and corrected. Waveforms and trends should always be assessed concurrently with changes in other monitored variables. Most of the newer CO_2 monitors either have waveform and trend display built in or allow integration with other monitors to offer these functions.

Capnogram waveforms are typically recorded or displayed at two different speeds: real time (high) speed at 12.5 mm/s, or trend (slow) speed at 25 mm/min (fig 4.1). At trend speed, slowly developing trends become more apparent, but finer details of each breath are lost. The slow speed capnogram displays each breath rising monotonously to the same or nearly the same end tidal value and falling to the zero baseline. Sudden changes can easily be seen over consecutive breaths from the CO_2 waveform, and the entire display gives more information on gradual changes. Changes in

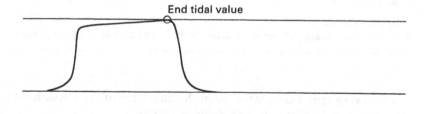

End tidal value

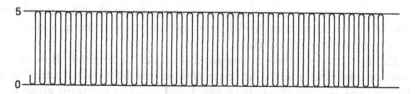

Fig 4.1 The two common types of capnographic waveforms: (top) real time (high) speed, 12.5 mm/second; (bottom) trend (slow) speed, 25 mm/minute

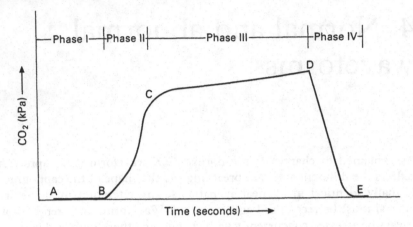

Fig 4.2 Recording of expired CO_2 concentration during a normal breath

---

Dead space

Anatomical dead space
- Caused by the conducting airways (trachea and bronchi)

Mechanical dead space
- Exists when patient is intubated
- Created by addition of non-gas exchanging tracheal tube and airway adapters

Alveolar dead space
- Areas of lungs are not perfused sufficiently to be capable of gas exchange

---

either the slow speed capnogram or in the fine detail of each breath inform of malfunctions in the patient or delivery system.

Carbon dioxide diffusion is a gradual continuous process at the alveolar level, but the CO_2 waveform has a definite shape because it illustrates the CO_2 concentration at the airway in both the inspired and expired gases. The idealised capnogram (fig 4.2) shows the theoretical changes in CO_2 with time over a respiratory cycle.

Expired gas or alveolar gas sampling can distinguish four distinct components: mechanical dead space, anatomical dead space, "ideal" alveolar gas, and alveolar dead space gas. Dead space is the term given to the parts of the airway that do not allow gas exchange (box). Gases from the mechanical and anatomical dead space are the first gases exhaled in series.

They contain almost no CO_2 because they have not been in the alveoli and consequently no gas exchange has taken place. Mechanical and anatomical dead space gases are seen as an initial flat portion or baseline of the capnogram (phase I). A raised phase I suggests rebreathing or a malfunctioning inspiratory or expiratory valve. The next exhaled gas, a mixture of gas from the anatomical dead space and gas from the alveoli, produces a rapid almost vertical rise in the capnogram tracing (phase II). This second phase corresponds with exhalation, but the initial flow of gas, which is composed mainly of gas from the conducting airways and not subject to gas exchange, does not induce a rise in the capnogram tracing. Phase III consists of a nearly horizontal plateau which coincides with exhalation of gas coming entirely from the alveoli. During exhalation the gas leaving the various parts of the lungs mixes thoroughly, so phase III reflects the average concentration of CO_2 from all the alveoli. The end of the alveolar plateau represents end tidal carbon dioxide concentration when the tension of CO_2 attains a fairly constant level; in the normal patient it is usually 0.2–0.4 kPa (2–3 mm Hg) lower than arterial CO_2. Because this expiratory "plateau" continues to rise slightly, it is probably more accurate to refer to it as phase III rather than as a plateau. Any slight upward slope at its end is normally attributed to gas exchange continuing during expiration in healthy subjects.

End tidal CO_2 very closely approximates the alveolar CO_2 concentration because it is measured when the patient exhales virtually pure alveolar gas. Gas from the alveoli and alveolar dead space are exhaled in parallel. If there is any appreciable alveolar dead space it is therefore impossible to sample or analyse ideal alveolar gas. Furthermore, the PCO_2 of end expiratory gas will invariably be less than that of "ideal" alveolar gas since it is diluted with gas from alveolar dead space.

The composition of inspired gas, alveolar volume, composition of mixed venous blood, and the flow rate at which gas enters the lungs are factors that can influence the rate of gas exchange and hence the slope of phase III. Measurement of the slope of phase III is a non-invasive method for estimating cardiac output, since mixed venous blood flow contributes to the generation of the slope.

The cycle begins again during phase IV, when the next inspiration starts. Fresh gases replace the alveolar gas at the sampling port and the capnogram shows a rapid fall in CO_2 towards baseline. As inspiration has started before the capnogram returns to zero, the first phase of respiration is traditionally referred to as the last phase. Inspiration concludes and the flow reverses, heralding Phase I again. The minimum concentration of CO_2 measured during the inspiratory phase, denoted by the horizontal line between the end of inspiration and the beginning of expiration of alveolar gas, is called the inspired CO_2 concentration (normally 0%). A flow signal is necessary to identify precisely the end of inspiration because expiration starts with the

57

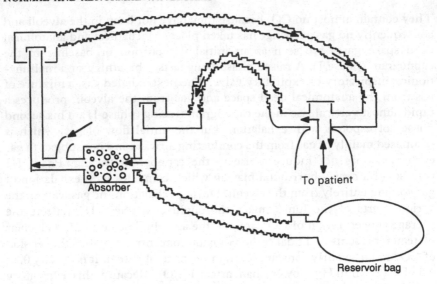

Fig 4.3 The circle anaesthetic breathing circuit

exhalation of gas from the anatomical dead space, which is identical to fresh gas, before the capnogram registers the arrival of alveolar gas.

The abnormal capnogram

As already stated, sudden changes in end tidal carbon dioxide concentration usually indicate changes either in circulation (for example, pulmonary embolism) or in ventilation (leak or disconnect) and can be easily seen "breath by breath" from the CO_2 waveform, while gradual changes are best seen from the trend display. Abnormalities in the capnogram must be analysed against the background of the patient's status and the pattern of ventilation accompanying the abnormality. If the patient is stable and time allows, spurious readings should be confirmed by arterial blood analysis of carbon dioxide and oxygen. To help find their cause, abnormalities in a capnogram can be classified by the phase in which they appear.

Phase I: the abnormal baseline

Failure of the capnogram to return to baseline usually implies the presence of carbon dioxide in the inspired gas. This can be inherent in the design of the breathing system (for example, Mapleson D or Jackson Rees) or it may be due to a malfunction when a circle anaesthetic system is used (fig 4.3).

58

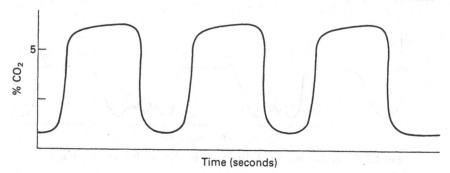

Fig 4.4 A raised baseline is a characteristic of Mapleson D or Bain systems

When the exhaled CO_2 is rebreathed, phase I of the capnogram fails to reach the zero baseline and may show a premature rise in CO_2 concentration during the inspiratory phase of ventilation, before the characteristic sharp upstroke associated with exhalation occurs. The $ETCO_2$ valve usually rises until a new equilibrium alveolar CO_2 concentration is reached at which excretion once again equals production. Partial rebreathing systems such as the Mapleson D and F (Jackson Rees's modification of Ayre's T piece) systems, commonly used in paediatric anaesthesia, permit valveless ventilation and have a characteristic capnogram (fig 4.4). Their use also removes the requirement for bulky CO_2 absorbers in the circuit and provides a lightweight, flexible circuit for ventilating small patients. The rebreathing design of these circuits also reduces water and heat loss from the airway during paediatric anaesthesia. The exact amount of rebreathing depends on many factors; exact estimation of end tidal carbon dioxide concentration and arterial PCO_2 is thus very difficult. The most reliable estimations are made in the presence of a normal alveolar plateau when a partial rebreathing circuit is in operation.

Inadequate fresh flow in valveless breathing systems can lead to rebreathing and the appearance of carbon dioxide in the inspired gas. For instance, during normal use of the Bain circuit in intermittent positive pressure ventilation with a Penlon Nuffield ventilator there is a non-zero baseline because the fresh gas flow is less than the minute volume (see fig 3.8). Capnometry gives qualitative but not quantitative information on rebreathing. Spontaneous ventilation with rebreathing of added dead space volume can also lead to an increased baseline.

Rebreathing should not occur with a circle anaesthesia system unless there is a malfunction in the system. Faulty check valves that allow bidirectional instead of unidirectional flow, the presence of a CO_2 absorber bypass circuit, or an improperly packed or saturated carbon dioxide

59

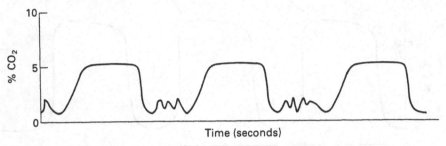

Fig 4.5 Raised baseline (with artefact caused by faulty expiratory valve)

absorber may cause an abnormally raised baseline (fig 4.5). Visual inspection of the check valves, bypass switch, and absorber will usually be required to detect these conditions.

A sudden increase in the baseline with an equivalent increase in the end tidal carbon dioxide concentration usually implies that there is contamination of the sample cell with water, dirt, or mucus.

Phase II: the slanted upstroke

The expiratory upstroke should be steep but becomes slanted and prolonged with obstruction to gas outflow or a slow rate of sampling. The expiratory time may be prolonged because of obstructive airways disease or asthma: carbon dioxide emerges too slowly from the lungs, producing a capnogram in which phase II becomes severely slanted and shortened and continues into Phase III. Alternatively, the obstruction to gas outflow may be external to the patient's airway—a kinked tracheal tube, for example. The delayed response of the sidestream analyser may prohibit the capnogram reaching zero during inspiration.

Phase III: the plateau

Ordinarily, the expiratory plateau should be flat and almost horizontal. The development of bumps and dips in the plateau with spontaneous respiratory efforts during mechanical ventilation often indicates light anaesthesia or inadequate muscle relaxation. Artefactual bumps, caused by pushing on the chest of an anaesthetised patient, which causes a little gas to move in and out of the lungs, are common during anaesthesia and surgical procedures. Biphasic respiratory plateaus occur when there is sequential rather than simultaneous emptying of the two lungs—a difference in compliance or resistance between lungs may cause one lung to expel gas later than the other, resulting in a wavering or uneven plateau. The expiratory plateau, together with the expiratory upstroke, slopes upwards when gas flow is obstructed, typically in patients with asthma or chronic obstructive lung disease.

Abnormally high plateau

Normal values for peak expiratory carbon dioxide partial pressure lie around 4.8 kPA (36 mm Hg). Higher CO_2 concentrations can be caused by physiological or pathological events. Deep sleep is associated with CO_2 concentrations in exhaled air of 5.8 kPa (44 mm Hg) or more as ventilation diminishes; the early increase in carbon dioxide observed by capnometry in malignant hyperpyrexia is an example of a pathological cause for a raised plateau.

Prolonged hypercapnia during anaesthesia is usually caused by too low alveolar minute ventilation (hypoventilation) due to inadequate tidal volume or respiration rate, or both. An interesting anomaly can occur during hypoventilation—the capnogram can display erroneously low values for CO_2. Hypoventilation in spontaneously breathing patients may be associated with significantly diminished expiratory gas flow rates, and if these decrease below the sampling flow rate, non-exhaled gases will be aspirated into the capnometer, lowering the displayed $ETCO_2$ concentration. This problem can be dealt with by reducing the sampling flow rate of the capnometer. In partial rebreathing systems, such as the Bain, prolonged hypercapnia may be due to inadequate flow of fresh gas causing too much rebreathing of CO_2. Noting that CO_2 does not fall close enough to zero in the inspiratory phase confirms this diagnosis. Leaks in the ventilator system should also be considered as a possible cause of hypercapnia, as should the absorption of CO_2 from the abdominal cavity during CO_2 laparoscopy.

Prolonged hypercapnia in the spontaneously breathing patient after general anaesthesia may be due to residual neuromuscular block or respiratory centre depression. Pain, especially after thoracoabdominal surgery, may cause splinting of the diaphragm, aggravating the hypercapnia.

Many elderly patients with chronic respiratory diseases tolerate raised carbon dioxide concentrations and it is usually not appropriate to alter such increased levels of carbon dioxide. A cause must be found and corrective action taken when end tidal carbon dioxide increases suddenly from a normal baseline, especially in fit healthy patients.

Sustained low CO_2 with normal plateau

Intentional or accidental lung hyperventilation often results in normal alveolar plateaus at CO_2 concentrations less than 4.5 kPa (34 mm Hg). The commonest cause (box) of continuous hypocapnia during anaesthesia is an

Causes for low but relatively normal plateau

- Inappropriately high alveolar minute ventilation
- Hyperventilation
- Decreased carbon dioxide production
- Decreased delivery of carbon dioxide to lungs (embolism or severe hypotension)
- Inadequate tidal volume setting on ventilator
- Mechanical artefacts in carbon dioxide analysis system

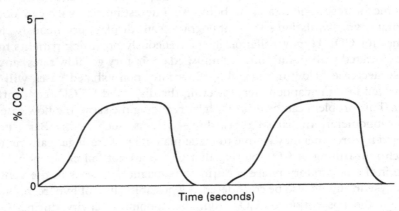

Fig 4.6 Decreased $ETCO_2$ with a normal alveolar plateau can occur in hyperventilation or in dead space ventilation. Comparison of $ETCO_2$ with $PaCO_2$ may distinguish these two conditions

inappropriately high alveolar minute ventilation or hyperventilation (fig 4.6). Hypocapnia in the spontaneously breathing patient after general anaesthesia may be a result of hyperventilation due to anxiety or pain.

Leaks in the sampling tube of a sidestream analyser can cause an appreciable reduction in the height of the plateau during intermittent positive pressure ventilation of a patient's lungs. The plateau is reduced because the capnometer aspirates room air together with exhaled gas, and the reduction in the plateau is proportional to the size of the leak. Absent or very low flows through the analyser-connector of the mainstream capnograph may lead to the development of artificial plateaus, which can erroneously suggest that expiration is continuing when in fact the patient is apnoeic.

In some circumstances the capnogram will show a low $ETCO_2$ with a widened alveolar-end tidal difference associated with a normal alveolar plateau. After the equipment malfunction or miscalibration is excluded, the most likely cause of the widened difference is a large physiological dead space ventilation. Chronic obstructive airway disease in adults and bronchopulmonary dysplasia in children, as well as pneumonia and other lung diseases, are commonly associated with large physiological dead space. Severe hypotension due to hypovolaemia will reduce pulmonary artery perfusion, increasing dead space ventilation, and will present as a normal plateau but low $ETCO_2$ with a widened alveolar-end tidal difference. Hypothermia and a reduction in the production of carbon dioxide are other causes of gradual falls in $ETCO_2$.

Abnormally low CO_2 with abnormal plateau

When the plateau is abnormal the $ETCO_2$ value cannot be taken as a good estimate of alveolar PCO_2 and hence certainly not as an estimate of arterial PCO_2. The absence of a normal alveolar plateau suggests that full and complete exhalation is not occurring before the next breath or that the patient's tidal volume is being diluted with fresh gas due to a small tidal volume, high aspirating sample rate, and high fresh gas dilution from the circuit.

Clinical signs and interventions will often help in diagnosis. Auscultation of the patient's chest may reveal ronchi, suggesting that small airway patency has been compromised by bronchospasm or secretions. Tracheal suctioning will often correct the partial obstruction and restore full exhalation. Bronchospasm may be treated by removing the cause or with bronchodilators or inhalational anaesthetics. The alveolar pattern is lost entirely when the tracheal tube is kinked, and a partial tracheal tube obstruction with cuff herniation into the lumen of the tube may interfere with full exhalation of tidal volume. Passing a suction catheter down the treacheal tube to determine patency will usually confirm or eliminate this possibility. In such cases, airway pressure will be raised or any attempt to manually ventilate the patient will meet high resistance.

Phase IV: the inspiratory downstroke

The inspiratory downstroke (phase IV) of the capnogram should be steep—it represents the replacement of carbon dioxide by fresh gas at the sampling site. Fresh gas aspirated into the sampling site of a sidestream analyser before the next inhalation has removed the end tidal gas of the previous sample can account for an abnormally slanted downstroke, and carbon dioxide in the inhaled gas or abnormally slow inspiration can produce a slanted and prolonged inspiratory downstroke. There may be

63

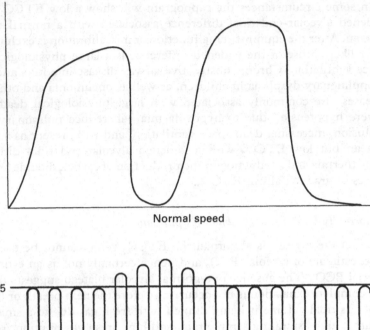

Normal speed

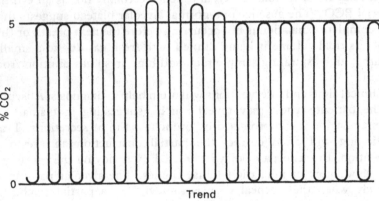

% CO_2

Trend

Fig 4.7 An acute but transient increase in $ETCO_2$ indicates a sudden increase in CO_2 delivery to the lung

carbon dioxide in the inhaled gas when the Bain system is used or when the inspiratory unidirectional valve of a circle system is malfunctioning.

Sudden increase in $ETCO_2$

A sudden increase in $ETCO_2$, which is easily identified in the trend mode (display at slow speed) may be caused by anything that acutely increases the delivery of carbon dioxide to the pulmonary circulation (fig 4.7). Such an increase most commonly occurs after the injection of sodium bicarbonate into the systemic circulation or after the release of a limb tourniquet or of a

64

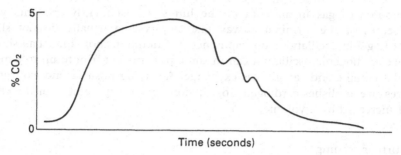

Fig 4.8 Normal capnogram with cardiogenic oscillations

vascular clamp from a major blood vessel. The chemical imbalance from injection of bicarbonate is short lived and of little clinical importance, whereas the rise after tourniquet release is more prolonged.

Sudden decrease in ETCO$_2$

Sudden loss of ETCO$_2$ to zero or near zero should immediately alert the anaesthetist to an interruption of ventilation. Possible causes include oesophageal intubation, ventilator malfunction or disconnection, or a totally occluded tracheal tube. An exponential decrease in the ETCO$_2$ trend usually indicates a serious cardiopulmonary event such as severe hypotension, cardiac arrest, or massive pulmonary embolus. The waveform is caused by the decrease in pulmonary perfusion and the resultant increase in the physiological dead space.

Cardiogenic oscillations

Cardiogenic oscillations, which are of no clinical importance, are undulations on exhalation synchronous with the electrocardiogram (fig 4.8). They are most often observed on the descending slopes of the capnogram when the expiratory phase is prolonged—during low respiration rates, for example, or if the tidal volume is too small. They may also occur at the end of a very long exhalation as the tidal flow of gas from the patient decreases to zero. These waveforms are more obvious when gas is sampled from deeper tracheal or bronchial areas.

Cardiogenic oscillations are caused by the heart beating against the lungs and diaphragm, causing transient flow at the end of the tracheal tube; they reflect changes in pulmonary blood volume. Blood drains from the pulmonary vascular system into the atrium and fills the ventricle during diastole, and the heart is thought to generate a small inspiratory movement which mainly affects the terminal bronchi of the lung, moving small

65

amounts of gas in and out of the lungs. Consequently, patients with diseases of the terminal airway like emphysema usually do not show cardiogenic oscillations on capnometry. Pneumotachography has shown that cardiogenic oscillation can in some patients contribute up to 8 ml to tidal volume and can aid gas exchange. Applying positive end expiratory pressure abolishes cardiogenic oscillations, probably because of the effect of increased lung volume.

Further reading

Swedlow D. Capnometry and capnography: the anesthesia disaster early warning system. *Seminars in Anesthesia* 1986;5:194–205.

5 Clinical importance of the abnormal capnogram

The decrease in deaths related to anaesthesia over the past 30 years has not deterred anaesthetists from continuing to audit mortality data. The information gathered is used to identify preventable causes of death in the hope of improving anaesthetic practice. It is difficult to construct or find reports of statistically valid studies showing which monitoring modalities prevent anaesthetic accidents. Prospective controlled trials of safety monitoring would not be ethically acceptable because of the need for an unmonitored control group, which could be considered to be at greater risk. Therefore most of the information on anaesthetic mortality has been collected retrospectively.

In recent years many reports have highlighted that problems with ventilation are the major cause of anaesthetic disasters. These have mainly arisen from an unrecognised oesophageal intubation, a disconnection in the breathing system, a kinked or dislodged tracheal tube or breathing system tubing, incorrect ventilator settings, or inadequate spontaneous or assisted ventilation during general anaesthesia.

Pressured by its medical malpractice insurer, Harvard Medical School appointed a risk management committee to investigate and suggest action on accidents related to anaesthesia between 1972 and 1985. Their audit revealed that most of the cases of major morbidity or death could have been prevented. Seven of the 11 anaesthesia related injuries in the case analysis at the nine hospitals affiliated with Harvard were associated with inadequate ventilation. The committee concluded that more meticulous monitoring would have given warnings early enough to allow the anaesthetist to correct the situation. The Harvard monitoring standards were instituted in response to this investigation.[1] The basis of the standards, which are the minimum, is the continuous monitoring of ventilation and circulation.

Closed claims analysis

Closed claims analysis of outcome of medical care is based on data from closed claims files of lawsuits. This method of audit is applicable to the analysis of anaesthetic morbidity and mortality data, but it is not a reliable estimate of true incidences of occurrence. Underreporting and lack of

67

participation by some insurance groups leads to underestimation of the true number of mishaps. There is also no control group for comparative analysis. Despite these limitations, closed claims analysis can provide exceedingly important information about anaesthetic outcome. In the American Society of Anesthesiologists' (ASA) closed claims study, respiratory events were responsible for the largest class of problems precipitating litigation. The ASA standards require oxygenation, ventilation, circulation, and temperature to be continually evaluated. While clinical assessment of oxygenation and ventilation is considered acceptable, the ASA standards encourage pulse oximetry and capnometry.

Analysis of the ASA closed claims database of serious intraoperative adverse respiratory events reveals that when adverse outcome followed the use of pulse oximetry or capnometry, either alone or in combination, the usual cause was the inappropriate use of monitoring equipment or the failure of the anaesthetist to notice or respond to critical monitored information. The commonest reasons cited for misuse of the monitoring equipment were that the alarms were disabled, accidentally or deliberately, and that the monitor was not turned on. The acknowledged high incidence of false alarms during the use of pulse oximetry and capnometry probably contributes to the dangerous practice of disabling alarms. In the closed claims study, no cases of intraoperative inadequate ventilation were reported when pulse oximetry was used. The analysis emphasises that the ASA standards for basic intraoperative monitoring can be effective only if the monitors are used correctly.

Standards for monitoring

Mandatory monitoring was introduced with the acknowledgement that a balance must be achieved between the requirement for extra monitoring and the possible benefits resulting from the resources allocated. The economic argument used in favour of monitoring suggests that making a relatively small financial commitment on monitoring equipment could save a great deal of money on malpractice insurance or on settlements following several injuries or deaths due to anaesthesia. To optimise the cost-benefit, a range of factors was considered before monitoring such as capnometry was made mandatory in the United States. Capital expense, ease of use, reliability, sensitivity, specificity, and distracting influence were among the factors considered. The standards had to be technically achievable since it would be futile to mandate capnographic monitoring based on, for example, Raman scattering so early in this technique's development.

The fundamental focus of the standards is intended to be behaviour rather than technology. Capnometry may be the most appropriate monitor for continuous monitoring of ventilation, but individual anaesthetists are

---

Guidelines on monitoring during routine anaesthesia

Variables monitored
- Oxygen supply
- Oxygenation of patient
- Airway
- Ventilation
- Circulation
- Temperature
- Depth of anaesthesia
- Neuromuscular function
- Inspired oxygen

Methods used
- Clinical monitoring
- Pulse oximetry
- Capnometry
- Stethoscopy
- Electrocardiogram
- Non-invasive blood pressure monitoring
- Ventilator alarms
- Oxygen analyser
- Postoperative anaesthesia care and monitoring

The standards are minimum requirements

---

not compelled to use any specific method. The standard is being observed and the desired early warning system is functioning appropriately when some form of continuous monitoring of ventilation, be it clinical or technological, is in operation.

The International Task Force on Anaesthesia Safety, a working party of the World Federation of Societies of Anaesthesiologists, has developed a set of international safety standards over the past few years; these were officially adopted at the tenth world congress of anaesthesiologists, held in the Netherlands in June 1992. The task force consisted of representatives from 10 countries and took a global view of anaesthesia safety. The adopted standards recommend basic, intermediate, and optimal requirements for giving anaesthesia, patient monitoring, and patient support (box).

Financial considerations, especially in Britain, have limited the more widespread use of monitoring technology, despite the fact that capnometry compares favourably with other discretionary costs associated with anaesthesia (fig 5.1). Cash starved health authorities may be slow to allocate funding for monitoring equipment in the absence of documented problems and in face of rigorous campaigning for funds from other specialities. In the United States it was argued that reductions in insurance premiums as a result of a decreased incidence of claims related to anaesthesia would offset the cost of increased monitoring requirements. The financial argument

69

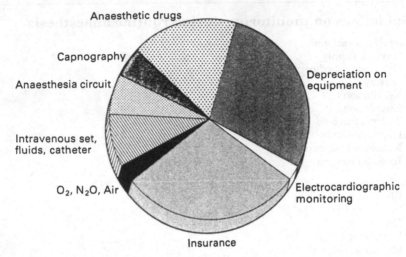

Fig 5.1 Cost of capnographic monitoring in relation to other discretionary costs associated with anaesthetic practice. (Reproduced with kind permission of D Swedlow, Seminars in Anaesthesia 1986;5:194–205)

now being used against further introduction of new technology is that anaesthesia is now so safe that future resource allocation would not be cost effective.

Against this background, capnometry has become a basic safety monitor primarily because carbon dioxide detected in a patient's exhaled gas gives evidence of ventilation. Since most major accidents involve ventilation failure, capnographic monitoring of ventilation should alert the practitioner to a problem with ventilation soon enough to allow a response that will prevent injury. Capnographic monitoring has become mandatory during routine anaesthesia in many countries, including the United States and Holland, and is increasingly used in Britain, following its recommendation in the monitoring guidelines published by the Association of Anaesthetists of Great Britain and Ireland in 1988.[3]

Capnometry enhances safety and provides a useful early warning for averting airway mishaps. The technique facilitates mechanical ventilation and contributes to the management of many medical conditions unrelated to anaesthesia. Abnormalities on the capnogram must be considered against the background of the patient's status and the pattern of ventilation accompanying the abnormality. If the patient is stable and time allows, spurious readings should be confirmed by analysis of carbon dioxide and oxygen from arterial blood. Sudden major changes in end tidal CO_2 are easily seen from the differences in consecutive waveforms; gradual changes are best seen from the trend display. Capnometry and pulse oximetry monitoring during general anaesthesia give valuable physiological and

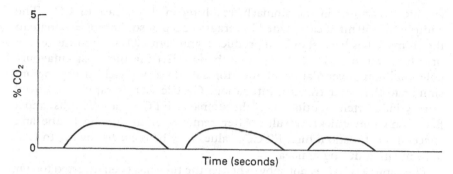

Fig 5.2 When an endotracheal tube is inadvertently placed in the oesophagus the normal capnogram is not seen. Small and transient capnograms may be due to gas entering the stomach during manual ventilation

safety data, and some important clinical situations where these monitors contribute are highlighted below.

Distinguishing tracheal from oesophageal intubation

The absence of ventilation that may result from accidental oesophageal intubation rapidly produces hypoxia and death if not quickly detected. Pre-oxygenation may delay the onset of the hypoxia and lead to a false sense of security, and the clinical diagnostic signs of inadvertent oesophageal intubation may all be misleading.

Ascertaining whether the tracheal tube has been properly inserted into the tracheal aperture is often difficult. During the intubation it may be difficult or impossible to observe the tip of the tracheal tube going into the larynx. The sign of chest movement supported by auscultation will not invariably establish correct tracheal intubation, especially in obese patients. Some anaesthetists palpate over the trachea just below the cricoid cartilage for the cuff of the tracheal tube by inflating and deflating the cuff repeatedly, but in one report palpation of the tracheal tube cuff in the neck proved misleading. Condensation of water vapour in the tube lumen, although less likely with oesophageal intubation, can occur and hence is not a reliable sign. Pulse oximetry may be slow to indicate the hypoxia that results from oesophageal intubation.

The most reliable test for the diagnosis of oesophageal intubation is the absence of carbon dioxide in the exhaled gas. A normal shaped capnogram confirms correct placement of the tracheal tube and ventilation during intubation. If the tube is inadvertently placed in the oesophagus, either some small transient capnograms are observed or none at all (fig 5.2). When some are seen, gas has entered the stomach during the initial manual ventilation, or certain antacids or carbonated beverages may have caused

71

chemical reactions in the stomach, resulting in the release of CO_2. The addition of 350 ml of carbonated beverages or 3 g of sodium bicarbonate to the stomach has been shown to produce significant CO_2 waveforms for six breaths or more after oesophageal intubation. Even in the most unfavourable conditions, ventilation of the stomach does not yield a capnogram similar to that after tracheal intubation. Gastric CO_2 is rapidly removed during inadvertent ventilation of the stomach, $ETCO_2$ rapidly falls, and a flat capnogram quickly results. After removal of the tracheal tube and successful extubation, high $ETCO_2$ values may be recorded owing to CO_2 accumulation during apnoea.

The capnogram does not show whether the tube has been inserted too far down the trachea and has entered a mainstem bronchus or has not been inserted far enough and lodges precariously at the larynx, where it is in danger of slipping out. The increased emphasis on patient safety brought about by the introduction of capnometry has led the American Society of Anesthesiologists to require anaesthesiologists to identify CO_2 in the exhaled gas to confirm correct tracheal tube placement as part of the minimal monitoring standard for basic intraoperative monitoring. Consequently, carbon dioxide analysers should be available in the operating theatres, intensive care, accident and emergency units, and resuscitation units or any site where tracheal intubation may be required.

Blind nasal intubation

Continuous capnographic recordings give valuable information about the position of the tracheal tube during the entire procedure of blind nasal intubation in the spontaneously breathing patient. If the CO_2 concentration is recorded at the proximal end of the nasotracheal tube, as the tracheal tube nears the opening of the larynx the end tidal concentrations increase. When the tube enters the trachea the typical flat topped capnograph waveform appears. If the tube is inadvertently entering into the oesophagus, smaller decreasing transient capnograms are seen. Ideally, capnometry should be used as an adjunct to clinical signs and other forms of monitoring during blind nasal intubation; occasionally it may have a key role when these normal clinical findings are difficult to interpret. Carbon dioxide monitoring can also be used to confirm correct placement and to detect dislodgement of double lumen tubes during thoracic procedures.

Cardiac arrest and cardiopulmonary resuscitation

As end tidal carbon dioxide concentration is partly determined by the amount of blood flow to the lungs (cardiac output), any reduction in pulmonary perfusion is reflected in sizeable reductions in the height of successive waveforms on the capnogram (fig 5.3). When cardiac arrest

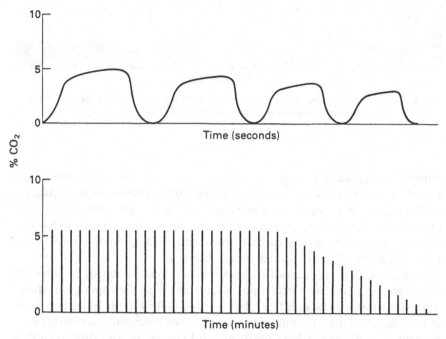

Fig 5.3 An exponential decrease (washout curve) in the height of successive capnograms is seen with a sudden disturbance in pulmonary perfusion such as cardiac arrest or sudden decrease in blood pressure

occurs, end tidal carbon dioxide concentration decreases dramatically. The capnogram disappears, signifying the absence of alveolar CO_2 due to collapse in pulmonary circulation.

End tidal CO_2 decreases during cardiac arrest, increases moderately with effective chest compressions, and increases significantly when the heart begins pumping effectively. Effective closed chest cardiopulmonary resuscitation can generate up to a third of normal cardiac output. Blood flow to the lungs, gas exchange, and the resulting $ETCO_2$ all remain lower than normal. Capnometry is useful for monitoring the effectiveness of cardiopulmonary resuscitation. Recovery of pulmonary perfusion can be confirmed by the rising capnogram. A simultaneous display of trend and the actual waveforms help evaluate the whole procedure and its efficacy.

The observation that rescuers performing chest compression tire when the capnogram deteriorates led to further studies investigating the role of CO_2 monitoring in cardiac arrest. In animal models CO_2 correlated significantly with myocardial perfusion pressures and cardiac output during cardiac arrest and resuscitation. End tidal CO_2 fell abruptly when ventricular fibrillation was induced in experimental models during survival

73

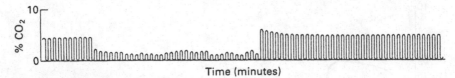

Fig 5.4 Trend capnogram depicting period of apnoea when mechanically ventilated patient was inadvertently disconnected from ventilator

studies, increased when cardiopulmonary resuscitation was started, and rapidly returned to normal pre-insult levels once spontaneous circulation was restored.[2] Studies have shown that CO_2 was significantly greater in animals that could be resuscitated than in those that could not. It has also been observed that during cardiac arrest and cardiopulmonary resuscitation CO_2 accumulates in mixed venous blood.

In human beings, successful resuscitation—restoration of an effective heartbeat (spontaneous circulation)—coincides with the capnogram showing a sudden steep rise in $ETCO_2$ because blood flow to the lungs increases, as does gas exchange. Capnometry therefore has potential as a non-invasive prognostic indicator of lung perfusion and cardiac output during cardiopulmonary resuscitation. The technique is already used in emergency medicine to assess changes in resuscitation protocols. Interpretation of vital clinical signs during cardiopulmonary resuscitation is unreliable, and examination of arterial blood gases is invasive and has not correlated with resuscitation and survival. The non-invasive method of capnometry continually assessing the efficacy of ongoing cardiopulmonary resuscitation would allow more rational patient management during cardiac arrest.

Apnoea

Capnometry is useful in the diagnosis of apnoea in patients breathing spontaneously or mechanically ventilated. If the end tidal carbon dioxide concentration waveform is initially normal but then suddenly drops to zero from one breath to the next, the most common cause is a ventilator disconnection. After a predetermined adjustable interval, usually 15 seconds, capnometers will sound an alarm when the next wave of carbon dioxide fails to arrive, indicating a respiratory arrest. A complete airway obstruction, caused for example by a fully kinked tracheal tube, is another possibility. If the cause is disconnection of the airway, apnoea is followed by an initially increased $ETCO_2$ when the integrity of the airway is reestablished (fig 5.4).

74

Positioning sampling tubes

- In sideport of disposable oxygen mask close to nose
- In nasal airways
- Attached to nasal prongs used to give oxygen
- Directly through a nostril

Although it is easier to monitor apnoea in intubated patients, sidestream capnometry can be used sucessfully in unintubated, spontaneously breathing patients. Sampling tubes can be positioned in various ways (box). One easily constructed modification of simple nasal cannulae involves the amputation of the tip of the cap of an ordinary needleless syringe, which is inserted through a small hole made with a needle at the base of the nasal cannula prongs. The syringe cap is then fully advanced into the nasal prong, where it completely obstructs the prong. The sampling tube from a capnometer can then be connected directly to the syringe cap and end tidal carbon dioxide can be accurately sampled during oxygenation. This approach may be advantageous in monitoring the ventilatory status of patients with chronic respiratory failure, in which excessive oxygen can cause CO_2 narcosis. Sampling tubes located external to the airway have the advantage of not being susceptible to the accumulation of secretions.

Capnometry in spontaneously breathing patients has greatly improved the safety of intravenous sedation techniques used for surgical procedures under local anaesthesia. Many of these patients are elderly and unsuitable for general anaesthesia. Clinical signs of respiratory difficulty may not be detected if surgical drapes cover the entire face during eye surgery or if there is poor lighting during microscopic procedures. Alveolar plateaus are often not observed when capnometry is used during spontaneous ventilation. Expired CO_2 cannot be used reliably to estimate $PaCO_2$ when a facemask is used because of the considerable entrainment of ambient air.

Capnometry can act as a disconnection monitor in unintubated, spontaneously breathing, anaesthetised patients. If a disconnection of the fresh gas flow occurs, rebreathing will quickly be evident on capnometry as the inspired CO_2 rises steadily. If the disconnection occurs between the breathing circuit and the patient, apnoea will be seen. The respiratory rate as measured by capnometry can be useful during spontaneous breathing as a crude indication of the depth of anaesthesia and fresh gas flow.

Monitoring adequacy of ventilation

Capnographic monitoring for apnoea is especially beneficial for patients in the recovery area after surgery. Postoperative patients are predisposed to respiratory failure because they have recently been given respiratory depressants as part of the general anaesthetic. Obese patients or those with pre-existing pulmonary disease who undergo upper abdominal and thoracic surgery are especially vulnerable. Pulse oximetry and CO_2 monitoring are excellent methods for detecting and managing apnoea or respiratory depression in this critical immediate postoperative period.

The capnogram provides useful continuous, non-invasive data about adequacy of ventilation during manipulation of ventilation in critically ill adults. For mechanically ventilated patients in postanaesthetic or intensive care units, CO_2 monitoring plays an important part in matching ventilation to metabolic activity. CO_2 monitoring allows continuous estimation of the ventilatory need of the patient non-invasively. These patients commonly have difficulty breathing, which may cause a ventilation-perfusion ratio imbalance that is reflected in the arterial to alveolar difference. Comparing arterial $PaCO_2$ with $ETCO_2$ gives information on the condition of the lungs; changes in arterial to alveolar difference need to be investigated further.

Capnometry is useful in patients who have ventilation abnormalities during mechanical ventilation. In bronchospasm, for example, the capnogram waveform will have a steeper slope throughout the expiratory phase and lose the normal alveolar plateau. The end tidal CO_2 concentration may be lower than the alveolar concentration and not accurately reflect PCO_2. Bronchodilator treatment can be evaluated by monitoring the slope of the expiratory phase of the capnogram. The slope will become less steep as the bronchospasm improves and the alveolar plateau better defined as a result of improved distribution of ventilation (fig 5.5).

The state of neuromuscular blockade in patients receiving muscle relaxants during mechanical ventilation can be shown by capnometry. Insufficient muscular relaxation allows the patient to "fight against the ventilator," and shallow, spontaneous clefts are seen in the alveolar plateau (fig 5.6). This type of capnogram is referred to as the "curare" capnogram. The negative clefts represent diaphragmatic contractions that are not associated with adequate inspiratory flow. Such ineffective diaphragmatic contractions send gas free of CO_2 past the sampling site, creating the cleft in the expiratory phase of the capnogram; the depth of the cleft is proportional to the degree of muscle paralysis. These dips or clefts may also signify hypoventilation with associated hypercapnia.

If neuromuscular blockade is incomplete, very strong stimuli may overpower the respiratory muscle weakness and cause small inspiratory efforts. The diaphragm is relatively resistant to neuromuscular relaxants and may recover function before the skeletal musculature, allowing capnometry to function as an early warning device for return of neuromuscular function during anaesthesia or ventilation in the intensive care setting. As muscular function improves, a chaotic pattern of respiration may develop if the patient cannot adequately support spontaneous efforts.

76

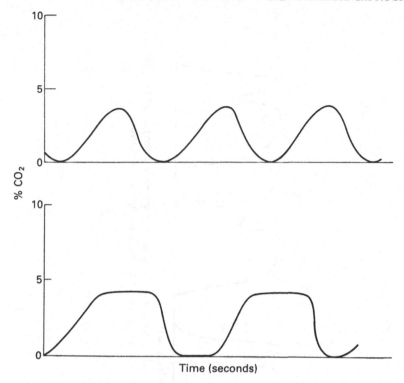

Fig 5.5 Monitoring efficacy of bronchodilator treatment with capnography: (top) obstructive pattern associated with bronchospasm; (bottom) improved plateau appearing after treatment

An inadequate sensitivity setting during mechanical ventilation can also be diagnosed by the capnogram. If the sensitivity is set too high, spontaneous respiratory efforts will generate a negative airway pressure that will fail to result in a ventilator assisted breath. The capnogram will display a dip or cleft during the alveolar plateau. A similar capnogram waveform is observed in patients with cervical transverse lesions.

The capnogram recorded during mechanical ventilation depends considerably on the mode of ventilation. Minute volume dividers such as the Manley ventilator produce a capnogram almost identical to the ideal. In contrast, ventilation with the Bain circuit produces a capnogram with a raised baseline because of rebreathing and a distorted inspiratory slope due to contamination of the fresh gas with expired gas (fig 5.7). When a coaxial system such as the Bain circuit is in operation the capnogram can assume different shapes. Apart from the patient's respiratory condition and carbon dioxide production status, the fresh gas flow rate, sampling point, and

77

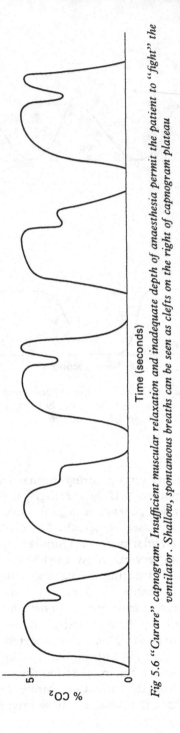

Fig 5.6 "Curare" capnogram. Insufficient muscular relaxation and inadequate depth of anaesthesia permit the patient to "fight" the ventilator. Shallow, spontaneous breaths can be seen as clefts on the right of capnogram plateau

Dips on the capnogram

- Result when diaphragmatic contractions are not associated with adequate inspiratory flow
- Depth is proportional to degree of muscle paralysis
- May signify hypoventilation with associated hypercapnia
- Appearance of dips signals return of neuromuscular function to diaphragm
- Dips during the alveolar plateau:
  Signal an inadequate sensitivity setting during mechanical ventilation
  Occur in patients with cervical transverse lesions

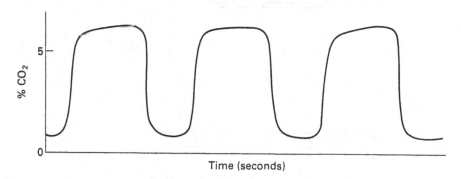

Fig 5.7 Raised baseline indicates a rebreathing of exhaled CO_2. This may be a normal feature of the anaesthesia delivery system (Mapleson D or Bain) or due to an exhausted CO_2 absorber in a circle system or a defective expiratory valve

volumes of the system should be considered when interpreting a capnogram when the Bain circuit is in operation. During spontaneous respiration, the Bain system requires a fresh gas flow of 2.5 times the minute volume to prevent rebreathing (fig 5.8).

Capnometry in intensive care

Carbon dioxide monitoring in the intensive care unit evaluates possibly more vitally important variables in critically ill patients than does any other single monitor. Patients in intensive care units often have multisystem failure and the availability of this non-invasive monitor aids in diagnosis and treatment. Many of these patients have alterations in ventilation which may be due to their illness, or they may be caused by therapeutic or diagnostic interventions. The most accurate measurement of monitoring ventilation in this setting is measuring $PaCO_2$ by sampling arterial blood, but this requires an arterial puncture, is costly, and currently provides only

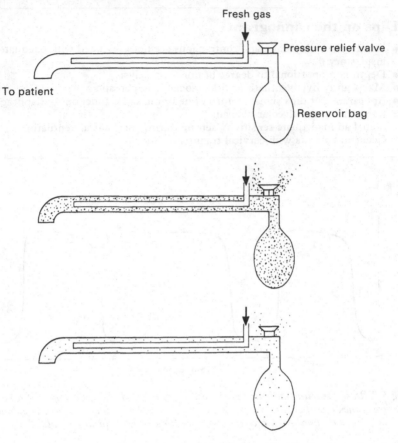

Fig 5.8 The Bain system. To prevent significant rebreathing the fresh gas flow for the system should be at least 2.5 times the minute volume. Stippling shows the distribution of carbon dioxide at end of expiration during low fresh gas flow (middle) and during high fresh gas glow (bottom). Arrow indicates entry of fresh gas to system

intermittent data. Capnometry and transcutaneous CO_2 monitoring continuously measure CO_2 without the requirement of repeated arterial puncture. Capnometry has several advantages over transcutaneous CO_2 monitoring (see chapter 3). Capnographic data provide information on ventilation-perfusion relationships. Dead space ventilation can also be estimated with capnometry in critically ill patients. Capnometry provides a measure of the end tidal partial pressure of CO_2, which is often a reliable reflection of $PaCO_2$.

Changes in $ETCO_2$ can indicate the critically ill patient's metabolic state. Acute increases in $ETCO_2$, reflecting important metabolic disturbances, may occur in critically ill patients. These changes may be seen in CO_2 waveform trends during rewarming after hypothermic cardiac bypass or after the sudden release of tourniquets from ischaemic areas. Monitoring the end tidal carbon dioxide concentration is important in these clinical situations, especially when the patient is paralysed or heavily sedated and not able to increase minute ventilation.

Patients who are alternating spontaneous and mechanical breaths during intermittent mandatory ventilation have a different capnograph pattern than those breathing spontaneously (fig 5.9). The initial pattern in the figure represents a fully controlled mechanical breath and is characterised by a stable, prolonged alveolar plateau. The following waveform shows a spontaneous breath, for which the duration of the alveolar plateau is much shorter. The $ETCO_2$ is the same as on mechanical ventilation despite the appearance of greater removal of CO_2 during the ventilated longer breath. The capnogram displays CO_2 concentrations over time and is independent of flow rate. The waveform does not reflect the volumes of gases exchanged. The final waveform in figure 5.9 illustrates a spontaneous breath occuring during a mechanical exhalation. If the expiratory valve is open, there is no obstruction to rebreathing from the expiratory limb; if there is no flow of fresh gases from the inspiratory line, then the patient may inhale a mixture of both fresh and exhaled gases. The $ETCO_2$ in this situation will not return to atmospheric baseline before the next breath begins.

Errors can result from passive CO_2 monitoring during low tidal volume high frequency ventilation for two reasons: the gas sampled may not be representative of alveolar gas as exhalation may not be fully completed due to the rapid respiratory rate, and capnometers may underestimate end tidal carbon dioxide concentration when their response time is greater than the respiratory cycle time of the patient.

Monitoring of end tidal carbon dioxide concentration is sometimes used to evaluate $PaCO_2$ during high frequency jet ventilation: $ETCO_2$ (squeeze $ETCO_2$) is measured after the delivery of a single large tidal volume during a respite from high frequency jet ventilation. The squeeze $ETCO_2$ is more representative of $ETCO_2$ than the passive $ETCO_2$. Arterial carbon dioxide is measured simultaneously to determine the arterial-end tidal difference and the $ETCO_2$ is used subsequently.

The $P(a\text{-}ET)CO_2$ difference can be used to determine the ideal level of positive end expiratory pressure for optimal oxygenation in critically ill patients with significant intrapulmonary shunting. Animal research shows that the level of positive end expiratory pressure that produces the lowest $P(a\text{-}ET)CO_2$ difference is the level at which optimal alveolar recruitment occurs. If the positive end expiratory pressure is raised above this level, significant alveolar overdistension and widening of the CO_2 difference occurs and oxygenation worsens. A low $P(a\text{-}ET)CO_2$ difference implies the optimal ventilation-perfusion relation, and this value gleaned from capnometry is central in the management of ventilated patients.

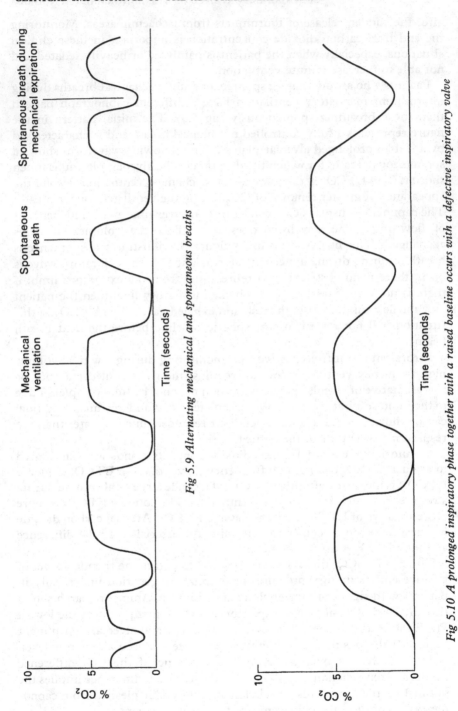

Fig 5.9 Alternating mechanical and spontaneous breaths

Fig 5.10 A prolonged inspiratory phase together with a raised baseline occurs with a defective inspiratory valve

Similarly, in mechanical ventilation with high airway pressures, increased dead space ventilation may occur by overextending alveoli. The alveolar distension can result in blockage or compression of nearby pulmonary capillaries. The resulting interruption of blood flow inhibits gas exchange. These clinical disturbances result in increased alveolar dead space and in dead space ventilation because ventilation continues despite decreased or non-existent perfusion through affected areas. With dead space ventilation, ventilation-perfusion ratios are high since the amount of ventilation far surpasses the amount of perfusion through these areas. The arterial to alveolar differences in these cases can be much wider than normal since reduced blood flow through these areas does not allow for gas exchange to occur and exhaled CO_2 concentrations become lower than normal.

Mechanical ventilation is usually instituted at an initial tidal volume of 10–12 ml/kg with a respiratory rate of 10 breaths/min. These settings should approximate normocapnia and if end tidal carbon dioxide is monitored arterial puncture should not be necessary. A cause should be found if large adjustments to settings are required to maintain the desired level. Possible explanations include hypermetabolism, altered alveolar dead space, functional or anatomical shunt, and capnometer malfunction.

Capnometry has a limited role during weaning from artificial ventilation. It can provide information about the adequacy of spontaneous respiration during weaning. Rising $ETCO_2$ values associated with an increasing respiratory rate and a loss of plateau on the capnogram signal indicate potential failure of weaning. Acute exacerbation of bronchospasm, which can develop during attempts at weaning, is indicated by a slow expiratory upstroke on the capnogram waveform. From the practical point of view, when CO_2 is monitored in patients with tracheostomies, the T connector should be placed directly on to the tracheal tube connector. Routine measurement of $PaCO_2$ is necessary for interpreting changes in the capnogram and $ETCO_2$ values during weaning.

Integrity of anaesthetic apparatus

Capnometry is an important early warning monitor for mishaps due to airway problems, leaks, and disconnections in the breathing circuit. Other means usually detect these hazards at a late stage, when the mishap starts to cause clinical problems in the patient. If ventilation fails after tracheal intubation, pulse oximetry will alert the clinician when oxygen saturation begins to fall, but this may have been delayed by preoxygenation. Capnometry would have indicated this problem at a much earlier stage, allowing the clinician vital extra time to rectify the situation.

Circuit leaks which decrease the minute volume may not be indicated by airway pressure monitoring but may be detected by CO_2 monitoring because the $ETCO_2$ gradually increases. Airway pressure monitors used to detect leaks in the breathing system occasionally fail to detect some disconnections. Under these circumstances a CO_2 monitor would detect disconnection instantaneously in mechanically ventilated patients. Carbon dioxide monitoring gives an early warning of CO_2 retention by the patient—due to a faulty Bain anaesthetic system, an exhausted CO_2 absorbent in a semiclosed anaesthetic system, leaks in the anaesthetic system, disconnections within the anaesthetic machine, and malfunction of valves in circle anaesthetic systems (fig 5.10).

If a normal $ETCO_2$ waveform drops to zero from one breath to the next the most common cause is a ventilator disconnection. A total airway obstruction—for example, caused by a fully kinked tracheal tube—or accidental extubation of the tracheal tube are other possibilities. A partially kinked or obstructed tube can result in increased or decreased $ETCO_2$, or in no change in $ETCO_2$, depending on the severity of the obstruction. Capnography is considered more valuable than capnometry in detecting partially obstructed tracheal tubes. With partial obstruction, full exhalation does not occur and the CO_2 waveforms display a prolonged phase II, a steeper phase III, and usually a lower plateau (fig 5.11). A deformed capnogram with a slowly rising leading edge is the typical waveform presentation of partial obstruction of the airway. These changes in waveform pattern occur earlier than changes in $ETCO_2$. However, tracheal tube obstruction must be severe (at least 50% occlusion) to produce changes in $ETCO_2$ or in the CO_2 waveforms. Other possible explanations for an airway obstructive type of capnograph waveform include foreign body, herniated tube cuff, and bronchospasm.

Coma

Head injury patients often have respiratory depression with associated raised arterial carbon dioxide blood tensions. There is a positive correlation between severity of head trauma as judged by the Glasgow coma scale and an early rise in arterial carbon dioxide tension. Monitoring ventilation and end tidal or arterial carbon dioxide is vital in the comatose patient, whatever the cause of the coma. To reduce intracranial pressure, ventilation is often finely adjusted to attain particular lower arterial CO_2 concentrations. During neurosurgical procedures, particularly with space occupying lesions, as part of the anaesthetic technique the arterial carbon dioxide tension is lowered to make use of the interaction between carbon dioxide and brain blood flow to control intracranial volume and pressure.

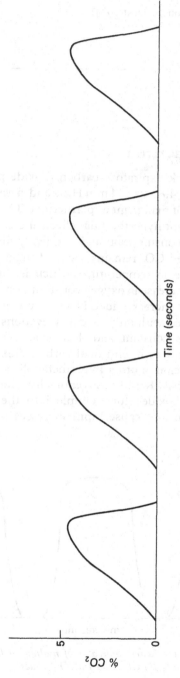

Fig 5.11 A slowly rising ascending limb on the capnogram may indicate a partial obstruction of the airway. This could be secondary to bronchial asthma, bronchospasm, mucus in airways, or a kinked endotracheal tube

---

Causes of airway obstructive type of waveform

- Partial airway obstruction (at least 50%)
- Foreign body
- Herniated tube cuff
- Bronchospasm

Malignant hyperthermia

Normal values for peak expiratory carbon dioxide partial pressure lie between 36 mm Hg and 4.8 kPa (44 mm Hg), and a rising $ETCO_2$ is an early sensitive indicator of malignant hyperthermia. The rapidly increasing metabolic rate in malignant hyperthermia reflects increased carbon dioxide production and oxygen consumption and is quickly displayed as a rising $ETCO_2$ while the inspired CO_2 remains constant (fig 5.12). Capnographic changes precede the rise in temperature, which is due to the chemical reactions caused by the increase in oxygen consumption and carbon dioxide production. Major rapid increases in $ETCO_2$ in malignant hyperthermia have often been associated with only modest elevations in temperature.

If ventilation remains constant and dead space ventilation does not change, unexpected increases in end tidal carbon dioxide tension should quickly lead to the suspicion of other hypermetabolic states such as severe sepsis or a thyrotoxic crisis. Rapid intravenous infusion of bicarbonate or insufflation of carbon dioxide—for example into the abdominal cavity during laparoscopy—can also cause rapid increases in end tidal carbon

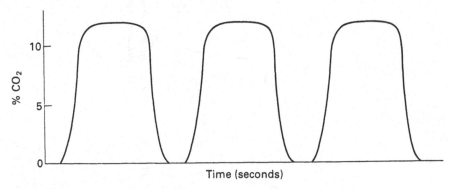

Fig 5.12 The capnogram is the fastest indicator of malignant hyperthermia, showing rising $ETCO_2$ and $PaCO_2$ values

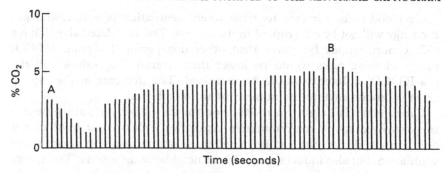

Fig 5.13 *An exponential decrease (from point A) in end tidal carbon dioxide concentration can be caused by severe hypotension, cardiac arrest, or pulmonary embolus. Point B shows the sudden increase in $ETCO_2$ after intravenous injection of sodium bicarbonate*

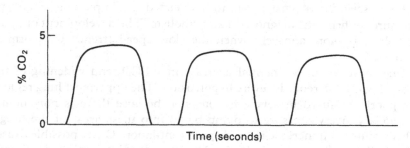

Fig 5.14 *A gradual fall in end tidal carbon dioxide concentration over several minutes suggests an increased minute ventilation, falling cardiac output, or a decrease in pulmonary perfusion*

dioxide (fig 5.13). These transient increases in exhaled carbon dioxide are most obvious when there is a constant rate of ventilation. A rapid but transient rise in CO_2 during laparoscopy may be an early sign of venous CO_2 embolism, which may occur after accidental insufflation of CO_2 into the venous system.

Pulmonary embolism

Pulmonary embolus (whether caused by thrombus, fat, or air) occludes blood flow through affected capillaries, resulting in a portion of blood flow that does not exchange gas or in dead space ventilation. If the situation is severe enough the overall CO_2 output from both lungs can be severely reduced; $ETCO_2$ is also reduced (fig 5.14). A pulmonary embolus may

87

cause blood to be directed to areas where ventilation is better, and gas exchange will not be interrupted in these areas. The ventilated alveoli have CO_2 concentrations far lower than other non-perfused alveoli. $ETCO_2$ values in these cases would be lower than arterial CO_2 values and the $P(a-ET)CO_2$ difference would be increased. The decrease in the capnogram is proportional to the number of alveoli affected.

Capnometry is helpful in monitoring for air embolism during neurosurgery. Blockage of the pulmonary capillaries with tiny air bubbles not only causes direct obstruction, leading to an increase in dead space ventilation, but also induces spasm in the neighbouring vessels. The spasm is aggravated by the increasing pressure in the air bubbles if nitrous oxide continues to be administered (due to the differing solubilities of these gases) and accounts for most of the air embolism effect. The typical capnograph presentation of an air embolism is similar to that of a pulmonary thromboembolism—the $ETCO_2$ progressively decreases and pulmonary artery pressure and right atrial pressure rise and there is a decrease in systemic blood pressure. End tidal CO_2 would be expected to fall with severe hypotension, but the associated rise in pulmonary arterial pressure confirms the diagnosis of air embolism. The development of an air embolism is more apparent when the slow speed (trend) waveform is displayed.

One cause of an exponential decrease in $ETCO_2$ and widening of the $P(a-ET)CO_2$ difference is severe hypotension. The uppermost lung regions are poorly perfused in severe hypotension because there is not enough pulmonary artery pressure to pump blood into these areas of the lungs. This results in an increase in dead space ventilation. Other possible causes of a similar capnogram include circulatory arrest with continued pulmonary ventilation and cardiopulmonary bypass. Hypocarbia is a feature of hypothermic cardiopulmonary bypass because most of the cardiac output during extracorporeal circulation bypasses the lungs and hypothermia is associated with a decrease in carbon dioxide production. The sampling catheter of the capnometer can be connected to the gas venting exit of the oxygenator chamber to assess the CO_2 tension of the circulating blood.

1 American Society of Anesthesiologists. *Standards for basic intraoperative monitoring, House of Delegates of the American Society of Anesthesiologists, October 21, 1986*. Park Ridge, IL: ASA, 1986.
2 Sanders A. Capnometry in emergency medicine. *Annals of Emergency Medicine* 1989; **18**: 1287–90.
3 Association of Anaesthetists of Great Britain and Ireland. *Recommendations for standards of monitoring during anaesthesia and recovery*. London: The Association, 1988.

Further reading

Carlon G, Cole R, Miodownik S, Kopek I, Groeger I. Capnography in mechanically ventilated patients. *Critical Care Medicine* 1988;**16**:550–6.

Duncan P, Cohen M. Pulse oximetry and capnography in anaesthetic practice: an epidemiological appraisal. *Canadian Journal of Anaesthesia* 1991;**38**:619–25.

Eichhorn J. Monitoring standards for clinical practice. *International Anesthesia Research Society: Review Course Lectures.* 1988;**1**:113–9.

Lunn JN, Devlin HB. Lessons from the confidential inquiry into preoperative deaths in three NHS regions. *Lancet* 1987;**ii**:1384.

Paloheimo M, Valli M, Ahjopalo H. *A guide to CO_2 monitoring.* Helsinki: Datex Instrumentarium, 1988.

Raemer DB, Philip JH. Monitoring anesthetic and respiratory gases. In: Blitt CD, ed. *Monitoring in anesthesia and critical care medicine.* 2nd ed. New York: Churchill Livingstone, 1990:373–86.

Szaflarski N, Cohen N. Use of capnography in critically ill adults. *Heart and Lung: The Journal of Critical Care* 1991;**20**:363–72.

Tinker JH, Dull DL, Caplan RA, Ward RJ, Cheney SW. Role of monitoring devices in prevention of anesthetic mishaps: a closed claims analysis. *Anesthesiology* 1989;**71**:541–6.

Zelcer J. Monitoring and anaesthetic equipment: standards and safety. *Current Opinion in Anaesthesiology* 1992;**5**:812–5.

6 Capnography in paediatric anaesthesia

Monitoring respiratory gases in paediatric patients has its own problems. The small size and immature anatomy of the patient must be considered. Accurate capnometry was difficult until quite recently in small infants, who are commonly ventilated through partial rebreathing circuits, in which the ratio of tidal volume to fresh gas flow is very small. The exhaled gas is often diluted and the end tidal gas measurement underestimates the real value. The high respiration rates often used in paediatric ventilation make measuring end tidal carbon dioxide concentration, which reflects true alveolar concentration, quite difficult. There is also potential for gas leakage as paediatric tracheal tubes have no inflatable cuff owing to the narrowness of the trachea. Recent advances in monitoring technology, however, have provided several methods of obtaining accurate capnometry in small infants (box).

Since high ventilatory rates and the low tidal and minute volumes are common in paediatric anaesthesia, producing and interpreting the capnogram can be difficult. The expired gas needs to be collected without contamination and the sampling rate must be high enough not to distort the true waveform. If these conditions are not met artefactual defects in the capnogram may be misinterpreted as physiological changes or abnormalities. Alveolar gas can be sampled reliably from small children under proper conditions by using modern equipment with a very small volume sampling cell.

Obtaining accurate capnometry in small infants

- Small volume sampling cell
- Adapters for paediatric tracheal tubes to minimise dead space
- Special tracheal tubes with sampling catheters within the wall of the tube and port at distal end for gas monitoring
- Heat and moisture exchanger may increase accuracy of proximal measurements
- Optimal gas flow is 150 ml/min
- Transcutaneous monitoring has a relatively long response time and is most appropriate for trend analysis
- Transconjunctival gas monitoring has a relatively fast response time

---

In capnometry, the term accuracy refers to the difference between the measured end tidal CO_2 ($PETCO_2$) values and the true end tidal values. Because alveolar ventilation is a smaller fraction of the fresh gas flow in children than in adults, discrepancies between the true end tidal gas concentration and the sampled end tidal gas concentration may be clinically important. Although a small physiological difference may exist between $PETCO_2$ and arterial PCO_2 ($PaCO_2$) because of the effect of dead space ventilation, $PETCO_2$ usually approximates $PaCO_2$ in intubated children with normal cardiovascular and respiratory physiology on both continuous and single breath analysis. This correlation is preserved in those infants and children with acyanotic congenital heart disease or a pure left to right shunt. In those with right to left shunts, however, a combination of venous admixture and low pulmonary blood flow with resultant dead space ventilation can produce a significantly increased alveolar dead space, which can increase the difference between arterial and end tidal PCO_2. Increasing the venous admixture increases the difference by shunting venous blood to the left side of the heart, where the CO_2 tension is greater than in the alveoli. In cyanotic congenital heart disease, although the $P(a-ET)CO_2$ difference is non-linear with respect to the $PaCO_2$, it is linearly related to the arterial oxygen saturation. A decrease in saturation of 10% caused by right to left shunting can be expected to increase the $P(a-ET)CO_2$ difference by 0.4 kPa on average.

In paediatric patients carbon dioxide is usually monitored with infrared spectography or mass spectrometry. Gas for analysis may be aspirated from the airway in sidestream analysers or analysed with mainstream infrared capnometry as it flows through a sensor inserted into the breathing circuit. Gas can be sampled from any site in the breathing circuit or tracheal tube.

Because infants have small tidal volumes, special paediatric adapters for use with paediatric tracheal tubes are available to minimise dead space (Datex, Helsinki, Finland). The adapter must have an internal diameter similar to that of the tracheal tube (2.5–4.0 mm). The special connector with sample port replaces the connector provided by the manufacturer with the tracheal tube.

Sampling in infants

Proximal end tidal CO_2 sampling refers to measurements between the breathing circuit and the tracheal tube and is suitable for infants and children weighing more than 12 kg ventilated with a volume ventilator and a Mapleson D partial rebreathing circuit. Proximal sampling is appropriate for infants and children of all weights ventilated with a paediatric circle breathing system, a Siemens-Elema Servo Ventilator or a Sechrist Infant Ventilator (Sechrist Industries, Anaheim, California) and a Mapleson D circuit. End tidal gas may be aspirated from the tracheal tube with a

91

23 gauge needle and 3 ml syringe at end expiration while the fresh gas flow is disconnected temporarily from the anaesthetic machine. The gas sample can be analysed for CO_2 in a blood gas analyser if a capnometer is not available. Single breath, end tidal carbon dioxide measurements from the proximal end of the tracheal tube accurately estimate $PaCO_2$ in infants and children and are useful when continuous capnometry is not available or when arterial cannulation is not indicated.[1]

In infants weighing less than 12 kg who are ventilated with a volume ventilator and a Mapleson D circuit, to obtain accurate end tidal CO_2 measurements respiratory gas must be aspirated from the tracheal tube (distal sampling). The sampling tube may be positioned at several sites within the tube, including the distal end or middle of the tube. For accurate measurement, the distance between the tip of the sampling catheter and the top of the tracheal tube should be less than 12.5 cm. Special tracheal tubes (Portex, Berck sur Mer, France), which have a sampling catheter within the wall of the tube and a port at the distal end for gas monitoring, minimise dead space. When the Mapleson F breathing circuit (Jackson-Rees's modification of Ayre's T piece) is used, these modified tracheal tubes also reduce the possibility of fresh gas contaminating the sample, which may otherwise artificially lower the end tidal CO_2 concentrations. These devices may provide accurate measurements, but one of their limitations is that if the sampling catheter becomes obstructed with secretions it may be necessary to increase or reverse the flow and flush the secretions back into the trachea, or to change tubes. Partial obstruction of the sampling tube by contaminants increases flow resistance, thus affecting the accuracy of the CO_2 measurement.

Proximal and distal sampling

In paediatrics, replacement of the tracheal tube may be a problem if intubation was difficult or if the airway is inaccessible, and the capnometer sampling tube should be changed before the procedure is started. Sampling with a distal catheter may be impossible if the purge mode of the capnometer is actuated in response to increased resistance to sample flow. The accuracy of measuring proximal aspirations is improved by inserting a small heat and moisture exchanger between the fresh gas flow inlet and the tracheal tube connector. The use of a humidifier, unfortunately, may lead to coalescence of water aerosols within the sampling line, which could obstruct the sampling line or damage the capnograph. The use of water filters, traps, and water permeable sampling lines can decrease the accuracy of some CO_2 analysers. Because water is removed by these water permeable catheters, some manufacturers have recommended that the vapour pressure of water (47 mm Hg) be ignored in the internal calibration. However, if the display value of CO_2 is internally calculated as dry gas (fractional

Proximal versus distal sampling

Advantages of proximal sampling
- Easier
- Less risk of obstruction
- Less risk of contamination
- Disconnection at insertion site less likely than with sidestream analysers

Advantages of distal sampling
- Higher measurements
- More accurate estimation of $PaCO_2$

concentration of CO_2 multiplied by barometric pressure) rather than wet gas (fractional concentration of CO_2 multiplied by barometric pressure minus 6.2 kPa (47 mm Hg)), considerable error may be introduced.

The sample flow rate for aspirating capnometry in paediatric anaesthesia is similar to that used in adults. Although the monitoring equipment has the capability for a wide range of gas flows, a gas flow of 150 ml/min is considered optimal as the capnographs generated are reliable in both children and adults and these flow rates do not affect the adequacy of spontaneous or mechanical ventilation. Even in very small infants these flow rates will aspirate only minimal amounts from the tidal volume and they will also allow accuracy at higher respiratory rates. In an infant weighing 3 kg with a respiration rate of 30 breaths/min, the total volume of gas sampled by a sidestream capnometer with 150 ml/min sample flow is 5 ml per breath. The typical tidal volume of 45 ml (15 ml × 3) for this weight of infant illustrates the relatively small percentage of the tidal volume consumed. Using higher gas flow when monitoring very small patients produces sinusoid deformities in the capnograph curves, which make interpretation difficult.

Non-invasive monitoring

Spontaneously breathing non-intubated children may be monitored with aspirating capnometry by taping the sampling catheter adjacent to the external nares. The Luer connector should not be inserted into a nostril as this may cause hypoxia if the other nostril becomes blocked. The large dead space of the face mask and high fresh gas flows make the $P(a\text{-}ET)CO_2$ value unreliable in these circumstances, but this safe, non-invasive technique may be used to facilitate continuous measurement of CO_2, to monitor respiratory patterns, and to detect airway obstruction in spontaneously breathing children.

93

Accuracy of infrared aspiration capnometry decreases when high respiratory rates are used during controlled ventilation. Accuracy can be improved by increasing the sample flow rate if very high respiratory frequencies are necessary. Monitoring at such high respiration rates is not often needed, but when it is, monitoring CO_2 in the airway provides continuous and non-invasive monitoring of apnoea, integrity of the anaesthesia system, and the respiration pattern, as well as showing as CO_2 trends. Reliable estimates of alveolar ventilation from the end tidal values are lost, as is the alveolar plateau on the capnogram. Gentle manual compression of the chest during expiration usually produces a normal plateau, and the end tidal CO_2 concentration measured will closely reflect the alveolar concentration of CO_2 in rapidly ventilating infants. In any case, the appearance of a plateau on the capnographic waveform does not always indicate that the end tidal CO_2 value reliably reflects the arterial CO_2 concentration. A flat alveolar plateau associated with a large arterial to end tidal difference may occur in patients with increased dead space ventilation (for example, bronchopulmonary dysplasia) and in small infants when proximal sampling is used inappropriately.

Waveform interpretation

The normal paediatric capnogram, like the waveform in adults, features a sharp upstroke as alveolar gas displaces tracheal gas and a slightly ascending plateau phase. Distortions of the normal shape suggest an abnormality in the patient, the breathing circuit, or the sampling technique. Ascertaining whether the tracheal tube has been properly inserted into the tracheal aperture is often difficult in infants. The most reliable test of oesophageal intubation is the absence of carbon dioxide in the exhaled gas. Oesophageal intubation is diagnosed by the absence of a waveform or small transient capnograms of decreasing height. As the stomach could contain some CO_2 after a mask induction or if certain antacids or carbonated beverages had been ingested, more than one breath should be observed to confirm tracheal tube placement. During inadvertent ventilation of the stomach, gastric CO_2 is rapidly removed, $ETCO_2$ rapidly falls, and a flat, square shaped capnogram quickly results.

If the tracheal tube has been inserted too far down the trachea and has entered a mainstem bronchus, end tidal measurements may initially be normal or slightly decreased. If, however, the bronchial placement is allowed to persist and the lungs are hypoventilated, end tidal PCO_2 may increase. A sudden disappearance of the capnogram indicates complete obstruction, disconnection of the breathing circuit, or a significant sudden increase in alveolar dead space caused by a major pulmonary embolus or cardiac arrest. A more gradual decrease in the end tidal CO_2 may reflect alveolar hyperventilation or pulmonary hypoperfusion. End tidal CO_2 may

Interpreting end tidal CO_2 measurements

Increase after being normal or slightly decreased
- Tracheal tube has entered mainstem bronchus

Sudden disappearance of capnogram
- Complete obstruction
- Disconnection of breathing circuit
- Major pulmonary embolus
- Cardiac arrest

Gradual decrease
- Alveolar hyperventilation
- Pulmonary hypoperfusion

Decrease to low but recordable value
- Excessive leak around tracheal tube

Slow increase
- Alveolar hypoventilation
- Increased CO_2 production
- In patient ventilated with circle circuit: faulty inspiratory or expiratory valves

Rapid and large increase
- Malignant hyperthermia

Sloping expiratory upstroke
- Kinked tracheal tube
- Airway obstruction

decrease to a low but recordable value in the presence of an excessive leak around the tracheal tube where the fresh gas flow replaces alveolar gas, thereby diluting the exhaled gas. A slowly increasing end tidal CO_2 value may occur with alveolar hypoventilation or in circumstances in which the CO_2 production increases. A very rapid and large increase in the end tidal CO_2 may suggest malignant hyperthermia. A gradual rise in both inspired and end tidal CO_2 in a patient ventilated with a circle circuit may indicate faulty inspiratory or expiratory valves. A sloping expiratory upstroke on the capnogram may indicate a kinked tracheal tube or airway obstruction (for example, bronchospasm or secretions). Cardiac oscillations occur in children with slow respiratory rates. They can be differentiated from inadequate muscle relaxant clefts by comparing heart rates to the frequency of waveform oscillation.

Baselines

In children ventilated through a semiclosed circle system with a CO_2 absorber the normal capnographic waveform shows a return to zero

Causes of raised baselines

- True rebreathing
- Diffusion of CO_2 in sampling lines
- Slow response of capnometer to rapid ventilatory rates in children

baseline. A raised baseline on the paediatric capnogram can be caused by true rebreathing or axial diffusion of CO_2 as it transverses the long sampling lines used in aspiration capnometry (fig 5.7). It is commonly recognised that in many partial rebreathing circuits used in paediatric anaesthetic practice the baseline will be neither zero or flat.

Rebreathing can occur in infants and children ventilated through Mapleson D breathing systems. The problem facing the anaesthetist is to differentiate between true rebreathing and distortion (artefact) of the capnogram. Boluses of CO_2 mix in the long sampling catheters used in paediatric patients with rapid ventilatory rates, thereby making the capnographic waveform misrepresentative. The normal waveform is replaced by one with a poorly defined upstroke and downstroke and a raised baseline. It has been suggested that parabolic distortion rather than molecular axial diffusion of the CO_2 plug occurs as it traverses the long sampling catheter.[2] Axial diffusion occurs across the fronts of CO_2 boluses travelling along the sampling catheter. Parabolic distortion may be caused by mechanical aspects of laminar flow where slower movement of gas occurs near the catheter wall due to increased friction. This distortion may produce partial fusion of CO_2 plugs and affect the interpretation of the $ETCO_2$ at rapid respiratory frequencies.

A raised baseline may also occur as a result of the capnometer's slow response time. The response times of the capnometer are transit time, rise (response) time, and total delay time. Transit time is the time taken for the sample to be transferred from the sampling port along the catheter to the analyser; it depends on the length and the diameter of the catheter, the rate at which the gas is aspirated, and the viscosity of the sample. The rise time, defined as the time it takes for the analyser output to change from 10% of the final value to 90% of the final value after the gas has entered the analysis cell, depends on the size of the sample chamber and the gas flow and is not affected by the length of the sample catheter. Rise time is a key determinant of a capnograph's ability to accurately reflect changes in CO_2 at the airway. The efficiency of the electronics and signal processing software also determines the duration of the rise time. Total delay time is the rise time plus transit time. Total delay time must be shorter than the respiratory cycle (time required for one breath) or the analyser will give spuriously low end tidal CO_2 and high baseline CO_2 readings.

96

A sustained low $ETCO_2$ with loss of normal alveolar plateau usually suggests partial airway blockage with delayed or incomplete emptying. An alternative cause in small children, who have characteristically small tidal volumes, is that the sampling flow rate may exceed the expiratory flow rate near the end of exhalation. In this situation the aspirating sample will be diluted with fresh gas from the anaesthetic breathing circuit. The result will be a drop off of the plateau and a fall of $ETCO_2$ secondary to dilution with fresh gas free of CO_2.

Reducing the flow rate of fresh gas temporarily or permanently and moving the sampling site closer to the tracheal tube connector will usually restore an alveolar plateau and result in a rise in the measured $ETCO_2$. Such contamination of end tidal gas with fresh gas flow may occur when T piece breathing circuits based on the Ayre principle, such as the Bain and Jackson Rees breathing circuits, are used. This contamination can be reduced by interposing a right angled connector between the breathing circuit and tracheal tube and inserting the capnometer sampling tube on the patient's side of the angle piece.

In very small newborns a sample rate of 100–250 ml/min may be too large to result in normal plateaus, even with reduction of fresh gas flow and movement of the sample port on to or within the tracheal tube itself. Capnometry in such a case can be used only as a monitor of cardiopulmonary status and of integrity of the anaesthesia breathing circuit

Transcutaneous CO_2 monitoring

Transcutaneous monitoring of carbon dioxide has been extensively used in neonatal intensive care units. The technique is most accurate in subjects with thin skin, such as neonates, who show less variability than in adults. The technique is particularly appropriate for use in infants because local skin blood flow tends to be high and repeated withdrawals of arterial blood may cause anaemia. The devices allow monitoring where vascular access is a problem, as is often the case in paediatrics. As already outlined in chapter 3, transcutaneous and arterial PCO_2 are correlated in stable, healthy patients. However, the relation in the acute care setting in the presence of pulmonary disease is less clear. Although transcutaneous carbon dioxide generally increases proportionally to arterial PCO_2, in states of extremely low perfusion (shock) a disproportionate increase in transcutaneous PCO_2 occurs because of the ensuing local cutaneous hypoxia. When transcutaneous devices are used as trend monitors in the critical care setting, a rise or fall in transcutaneous CO_2 values indicates a significant change in the cardiopulmonary system as a unit, leaving the clinician to determine the cause of the trend more specifically. For example, a rise in transcutaneous CO_2 may imply worsened alveolar ventilation from a variety of causes ranging from ventilator malfunction to worsened ventilation-perfusion

97

Transcutaneous CO_2 monitoring in children

Advantages
- Spares infant from blood sampling
- Allows monitoring where vascular access is a problem
- In critical care setting, use as trend monitor shows change in cardiopulmonary system as a unit

Disadvantages
- Long response time
- Need to raise skin temperature; risk of skin burns and increase in local production of CO_2
- Direct arterial measurements needed for calibration

mismatch, but may just as plausibly imply decreased tissue perfusion either globally or locally.

One of the major disadvantages of transcutaneous monitoring in children is the relatively lengthy response time (3–15 minutes), which is influenced significantly by probe temperature. A sudden change in arterial PCO_2 may not be detectable quickly enough for a therapeutic response to be made. Skin temperature must be raised to obtain clinically useful transcutaneous PCO_2 values since the higher the probe temperature the faster the observed response time. Skin burns have sometimes resulted from prolonged application of these devices, and it is now recommended that devices are resited every four hours to avoid burns. The other disadvantage of local warming is that it also measures local production of CO_2. Another limitation of transcutaneous CO_2 monitoring is the requirement for direct arterial measurements to calibrate the transcutaneous instrument.

Transconjunctival gas monitoring

The new technique of transconjunctival oxygen monitoring attempts to minimise the problems associated with slow diffusion time of respiratory gases across skin. The Clark type electrode in the instrument designed to measure conjunctival PO_2 (Orange 1 Oxygen Monitoring System; Orange Medical Instruments, Costa Mesa, California) is embedded in a polymethylmethacrylate, ring shaped scleral conformer that is contoured to fit the eye. The conformer is placed under the upper and lower eyelids so that the electrode lies against the palpebral conjunctiva. These devices may be kept in place for 24 hours or longer in unanaesthetised patients. Over a range of arterial oxygen pressures, conjunctival PO_2 correlates positively with PaO_2. The response time to an abrupt change in inspired PO_2 is relatively fast, usually about a minute. The conjunctival pressure is uniformly lower than arterial pressure, with wide variability among patients. Patient specific calibration is required to estimate absolute PaO_2 and the technique, like transcutaneous monitoring, is more appropriate for trend analysis. Transconjunctival monitoring of carbon dioxide is not yet available commercially.

Summary

The combined use of pulse oximetry and capnometry in transcutaneous anaesthetic practice provides continuous monitoring of oxygenation, ventilation, and tissue perfusion. A diminishing capnographic waveform (gradually or suddenly, depending on the rate of cardiac collapse) along with failure of pulse oximetry may indicate a decreasing cardiac output in a child. The presence or absence of blood pressure and pulses will quickly confirm capnographic and oximetric evidence of cardiac output failure. Pulse oximetry reduces both the frequency and severity of hypoxaemia during anaesthesia in children. Pulse oximetry is more likely to detect episodes of hypoxaemia, but capnometry may provide an early warning of ventilation mishaps such as oesophageal intubation and disconnections in children that could result in hypoxaemia before the onset of arterial haemoglobin oxygen desaturation.

1 Bhavani Shankar K, Moseley H, Kuomar A, Delph Y. Capnometry and anaesthesia. *Can J Anaesth* 1992;**39**:617–32.
2 Brunner JX, Westenkow DR. How carbon dioxide rise time affects the accuracy of CO_2 measurement. *J Clin Monit* 1988;**4**:134–6.

Further reading

Hardwick M, Hutton P. Capnography: fundamentals of current clinical practice. *Current Anaesthesia and Critical Care* 1990;**1**:176–80.
Stokes M, Hughes D, Hutton P. Capnography in small subjects. *Br J Anaesthesia* 1986;**58**: 814.

Conclusion

Recent advances in medical technology have made possible rapid accurate measurement of exhaled carbon dioxide in a wide variety of clinical conditions. The relative compactness and ease of use of modern capnometers facilitates their routine use during most anaesthetic procedures. Clinical applications range from assessment of correct tracheal intubation after induction of anaesthesia, to analysis of efficacy of cardiopulmonary resuscitation during cardiac arrest. Though carbon dioxide measurement and display can nowadays be achieved easily, interpretation of these findings and the possible need for subsequent action requires an understanding of the basic principles of the measurement and related physiology.

Epidemiological and medicolegal studies suggest that anaesthetic practice is safer now than ever before. Most anaesthetists would agree that monitoring has played an important part in this improvement in safety, but this has proved difficult to confirm. Oximetry and capnography have been shown to be efficacious in controlled situations in detecting earlier the type of events judged to be most harmful during anaesthesia, and the literature is full of reports of their role as early warning monitors.

The main argument against further investment in monitoring is financial, with health care planners claiming that anaesthesia has now become so safe that investment in this service may lack cost effectiveness. It must be emphasised, however, that one of the reasons that anaesthesia has become so safe has been the introduction of monitoring standards coincident with the recent advances in monitoring technology. In an era of financial rectitude in health care, it would be short sighted to curtail the investment in technological progress by ignoring diagnostic information systems that could further contribute to perioperative patient care.

Appendix

Units of measurement

%   = volume percentage

$\left.\begin{array}{l} \text{mm Hg} \\ \text{kPa} \end{array}\right\}$ = partial pressure

1% = 1 kPa
1 kPa = 7.5 mm Hg
1 mm Hg = 0.13% (or kPa)

pH is a measurement of the hydrogen ion (H^+) concentration in a solution—that is, the acidity or alkalinity

Abbreviations

$PaCO_2$	pressure of carbon dioxide in arterial blood
PaO_2	pressure of oxygen in arterial blood
$PETCO_2$	pressure of carbon dioxide in end tidal sample
$PACO_2$	alveolar pressure of carbon dioxide
VA	alveolar ventilation
VCO_2	carbon dioxide production
$ETCO_2$	end tidal carbon dioxide concentration
$P(a\text{-}ET)CO_2$	alveolar-arterial carbon dioxide tension difference

Index

||||| || ||| ||||||| |||| |||||| |||
9 780727 907967